OCD Studies on Water

Pharmaceutical Residues in Freshwater

HAZARDS AND POLICY RESPONSES

This document, as well as any data and any map included herein, are without prejudice to the status of or sovereignty over any territory, to the delimitation of international frontiers and boundaries and to the name of any territory, city or area.

Please cite this publication as:
OECD (2019), *Pharmaceutical Residues in Freshwater: Hazards and Policy Responses*, OECD Studies on Water, OECD Publishing, Paris, *https://doi.org/10.1787/c936f42d-en*.

ISBN 978-92-64-77633-3 (print)
ISBN 978-92-64-97741-9 (pdf)

OECD Studies on Water
ISSN 2224-5073 (print)
ISSN 2224-5081 (online)

Co-published by IWA Publishing
Alliance House, 12 Caxton Street, London SW1H 0QS, UK
Tel: +44 (0)20 7654 5500, Fax: +44 (0)20 7654 5555
publications@iwap.co.uk
www.iwapublishing.com

ISBN: 9781789061819 (Paperback)
ISBN: 9781789061826 (eBook)

The statistical data for Israel are supplied by and under the responsibility of the relevant Israeli authorities. The use of such data by the OECD is without prejudice to the status of the Golan Heights, East Jerusalem and Israeli settlements in the West Bank under the terms of international law.

Photo credits: Cover © Marian Weyo/Shutterstock.com

Corrigenda to OECD publications may be found on line at: *www.oecd.org/about/publishing/corrigenda.htm.*

Foreword

Pharmaceuticals are an important element of medical and veterinary practice, and their beneficial effects on human and animal health, food production and economic welfare are widely acknowledged. However, an area where we lack a common understanding is what happens when these pharmaceuticals are constantly discharged into the environment, through pharmaceutical manufacturing, consumption and excretion, and improper disposal of unused or expired products.

Residues of pharmaceuticals, such as hormones, antidepressants and antibiotics, have been detected in surface water and groundwater across the globe. High levels of pharmaceutical residues have been found downstream of pharmaceutical manufacturing plants, and conventional wastewater treatment plants are not designed to remove pharmaceuticals in wastewater. Furthermore, veterinary pharmaceutical residues from agriculture and aquaculture can enter water bodies without any treatment.

Evidence is growing of the negative environmental impacts, with laboratory and field tests showing traces of oral contraceptives causing the feminisation of fish and amphibians, and residues of psychiatric drugs altering fish behaviour. The mis- and over-use of antibiotics is an important contributing factor of the emergence and spread of antimicrobial resistance - a global health crisis with the potential for enormous health, food security and economic consequences.

Unless adequate measures are taken to manage the risks, pharmaceutical residues will increasingly be released into the environment. Ageing populations, advances in healthcare, and intensification of meat and fish production is spurring the demand for pharmaceuticals worldwide. At the same time, the need for clean water will also increase, including treated wastewater for agriculture and high-purity water for manufacturing pharmaceuticals. Climate change is further reducing water availability in sufficient quantity and quality, as well as increasing the risk and spread of disease.

As the challenge of meeting the 2030 sustainable development goals on water and health progresses, emerging issues such as pharmaceuticals in the environment should be prevented from becoming future traditional environmental and human health threats. Health ministers of the G20 have declared to fight antimicrobial resistance. The OECD Council Recommendation on Water calls for Adherents to prevent, reduce and manage all sources of water pollution, in surface and ground waters and related coastal ecosystems, while paying attention to pollutants of emerging concern such pharmaceutical residues. Stakeholders of the Strategic Approach to International Chemicals Management adopted environmentally persistent pharmaceutical pollutants as an emerging policy issue, and agreed international cooperation is crucial to build awareness and promote action on the issue.

Pharmaceutical residues in freshwater is a challenge that must continue to be addressed in ways that take into account advances in our knowledge. This report contributes to this growing imperative. It stresses the need for a better understanding of the effects of pharmaceuticals in the environment, calls for greater international collaboration and accountability distribution, and suggests policy actions to prevent and remedy the problem across the pharmaceutical life cycle. A cross-sectoral response, with economic and regulatory drivers from central government is recommended to incentivise action by pharmaceutical companies, healthcare providers, veterinarians, farmers and food producers, wastewater utilities and the general public. A focus on preventive options, such as disease prevention and improved diagnostics, the sustainable design, manufacture and procurement of pharmaceuticals, and restrictions on the use of pharmaceuticals with high environmental risk, will deliver the most long-term and large-scale environmental benefits.

This timely report brings together the expertise and experience of the health and environment communities. I am confident that policymakers can find both inspiration and pragmatic support in this report, to translate ambition into action on improving health and protecting the environment.

Rodolfo Lacy

OECD Environment Director

Acknowledgements

This report was prepared by the OECD Environment Directorate, under the leadership of Director Rodolfo Lacy and the subsidiary Climate, Biodiversity and Water Division led by Simon Buckle. The project manager and author of the report is Hannah Leckie. The project was delivered under the supervision of Bob Diderich, Head of the Environment Health and Safety Division, and Xavier Leflaive, Water Team Leader.

The Secretariat gratefully acknowledges the contributions of the delegates of Working Party on Biodiversity, Water and Ecosystems and the Joint Meeting of the Chemical Committee and the Working Party on Chemicals, Pesticides and Biotechnology. The Secretariat is also grateful to the Ministry of Infrastructure and Water Management, Netherlands and the Ministry of Environment, Republic of Korea for their generous financial support in developing this report.

The author expresses gratitude to those who contributed to the preparation and development of this report through the provision of their time, expertise and experience. The report benefitted from insightful discussions and the technical expertise of participants at the OECD Workshop on "Managing Contaminants of Emerging Concern in Surface Waters: Scientific developments and cost-effective policy responses" held on 5 February 2018, Paris. Special appreciation is extended to: Sara Sahlin (formerly OECD Environment Directorate) who provided valuable research for the development of chapters 1 and 2 of this report; Florence Metz (Department of Political Science, University of Bern, Switzerland) who provided a scoping study and guidance on the development of this project; and Frithjof Laubinger (OECD Environment Directorate) who provided research on pharmaceutical waste management and collection programmes.

Appreciation is extended to those who contributed case studies, including: Josef Hoppichler (Federal Institute for Less-Favoured and Mountainous Areas, Austria); Henrik Søren Larsen (Ministry of Environment and Food, Environmental Protection Agency, Denmark); Olivier Gras (Ministry for the Ecological and Inclusive Transition, France); Park Tae-Jin (National Institute of Environmental Research, Korea); Marc L. de Rooy, Julia Hartmann, Ana Versteegh and Tialling Vlieg (Ministry of Infrastructure and Water Management, Netherlands); Paula Paíga, Virginia Cruz Fernandes, Ana P. Carvalho, Manuela Correia, Olga Freitas, Sónia Figueiredo, Valentina F. Domingues and Cristina Delerue-Matos (Polytechnic of Porto - School of Engineering (ISEP), Portugal); Kerstin Bly Joyce (Environmental Protection Agency, Sweden); Florian Thevenon (WaterLex International Secretariat, Switzerland); and Nick Haigh (formerly Department for Environment, Food and Rural Affairs, United Kingdom).

The author is thankful to colleagues and experts who provided valuable input and review comments on the report, in particular: Simon Buckle, Bob Diderich, Xavier Leflaive and Leon Van Der Wal (OECD Environment Directorate); Valerie Paris, Ruth Lopert and Michael Padget (Health Division, OECD Directorate for Employment, Labour and Social Affairs); Alistair Boxall (Department of Environment and Geography, University of York, United Kingdom); Armelle Hebert (formerly Environment & Health Department, Veolia Research & Innovation); Klaus Kümmerer (Institute of Sustainable and Environmental Chemistry, Leuphana University of Lüneburg, Germany); Florence Metz (Department of Political Science, University of Bern, Switzerland); Fiona Regan (School of Chemical Sciences, Dublin City University,

Ireland); Stéphanie Rinck-Pfeiffer (Global Water Research Coalition); Jason Snappe and Sam Maynard (AstraZeneca); Astrid Louise Wester (formerly Department of Public Health and Environmental Determinants of Health, World Health Organisation); and Annemarie van Wezel (Institute for Biodiversity and Ecosystem Dynamics, University of Amsterdam, the Netherlands).

Editorial guidance from Janine Treves, communications services from Catherine Bremer, Beth Del Bourgo, Sama Al Taher Cucci and Jane Kynaston, and administrative support from Ines Reale and Anna Rourke are gratefully acknowledged.

Table of contents

Tables

Figures

Boxes

Abbreviations

AMR	Antimicrobial resistance
AOP	Adverse outcome pathway
API	Active pharmaceutical ingredient
BAT	Best available techniques
BCF	Bioaccumulation factor
BEP	Best environmental practices
CECs	Contaminants of emerging concern
CEPA	Canadian Environmental Protection Act
CIP	Chemical investigations programme
E1	Oestrone
E2	17β- estradiol
EE2	17α- ethinylestradiol
EDA	Effect-directed analysis
EDC	Endocrine-disrupting chemicals
EIA	Environmental impact assessment
EPPs	Environmentally persistent pharmaceutical pollutants
EPR	Extended producer responsibility
ERA	Environmental risk assessment
EU	European Union
EQN	Environmental quality norm
EQS	Environmental quality standard
FDA	[U.S.] Food and Drug Administration
GAC	Granular activated carbon
GMP	Good manufacturing practice
LC-HSMS	Liquid chromatography–high resolution mass spectrometry
LNG	Levonogestrel

LoQ	Limit of quantification
MSC	Minimum selective concentration
NOAEL	No observed adverse effect level
NSAID	Non-steroidal anti-inflammatory drug
OTC	Over-the-counter
PAC	Powered activated carbon
PBT	Persistent, bioaccumulative and toxic
PEC	Predicted environmental concentration
PNEC	Predicted no effect concentration
POP	Persistent organic pollutants
PPCPs	Pharmaceuticals and personal care products
UK	United Kingdom
U.S.	United States of America
vPvB	Very persistent and very bioaccumulative
WASH	Water supply, sanitation and hygiene
WFD	[EU] Water Framework Directive
WHO	World Health Organisation
WSP	Water safety plan
WWTP	Wastewater treatment plant

Executive Summary

About 2 000 active pharmaceutical ingredients (APIs) are being administered worldwide in prescription medicines, non-prescription drugs and veterinary drugs, the residues of which are of increasing environmental concern as the number and density of humans and livestock requiring healthcare escalates.

Active pharmaceutical ingredients are found in surface waters, groundwater, drinking water, soil, manure, biota, sediment and the food chain. Although the contribution of each emission source varies across regions and types, the dominant sources of pharmaceuticals in the environment stem from untreated household wastewater and effluent from municipal wastewater treatment plants. Emissions from manufacturing plants and intensive agriculture and aquaculture can be important pollution hotspots locally.

Because pharmaceuticals are intentionally designed to interact with living organisms at low doses, even low concentrations in the environment can have unintended, negative impacts on freshwater ecosystems. For example, active substances in oral contraceptives have caused the feminisation of fish and amphibians; psychiatric drugs, such as fluoxetine, alter fish behaviour making them less risk-averse and vulnerable to predators; and the over-use and discharge of antibiotics to water bodies exacerbates the problem of antimicrobial resistance – declared by the World Health Organisation as an urgent, global health crisis that is projected to cause more deaths globally than cancer by 2050.

Advances in monitoring can help close the knowledge gap and support policy responses

Most OECD countries have established watch-lists and voluntary monitoring programmes for certain pharmaceuticals in surface water, but the majority of APIs, metabolites and transformation products remain unmonitored and without ecotoxicity data. There are therefore a number of uncertainties associated with the environmental risk assessment of pharmaceuticals due to lack of knowledge concerning their fate in the environment and impact on ecosystems and human health, and the effects of mixtures of pharmaceuticals and other chemicals.

The cost of monitoring, limited data for policy development and an absence of a systematic approach to risk assessment were three barriers to taking action identified by governments in the 2017 OECD Questionnaire on Contaminants of Emerging Concern in Freshwaters. Advances in monitoring technologies and modelling can help close the knowledge gap and support policy responses. Real-time in-situ monitoring, passive sampling, biomonitoring, effects-based monitoring, non-target screening, hotspots monitoring, surrogate data methods, early-warning systems and holistic modelling can help identify and prioritise APIs in the environment, and anticipate sources of contamination. Country and international initiatives are crucial to improve the knowledge base and exchange of data, methodologies and technologies to address risks between countries and sectors.

Moving towards proactive policy action to curb pharmaceutical pollution

Current policy approaches to manage pharmaceutical residues are inadequate for the protection of water quality and freshwater ecosystems upon which healthy lives depend. They are often reactive (i.e. when risks are proven), substance-by-substance (i.e. individual environmental quality standards) and resource intensive. And diffuse pollution, particularly from livestock and aquaculture, remains largely unmonitored and unregulated.

All stakeholders along the pharmaceutical chain have a critical role to play in the transition to more effective management of pharmaceutical pollution. Voluntary participation alone will not deliver; economic and regulatory drivers from central government are needed.

Policy-makers will need to factor in financing measures for the upgrade, operating and maintenance costs of wastewater treatment plants, as well as policy transaction costs to facilitate the transition from reactive to proactive control of pharmaceutical residues in water bodies. The relative risk of pharmaceuticals should also be compared with other water pollutants (e.g. heavy metals, persistent organic pollutants and other contaminants of emerging concern) to achieve improvements in water quality and ecosystems in the most cost-effective way.

While acknowledging the critical role of pharmaceuticals for human and animal health, a combination of the following four, proactive strategies can cost-effectively manage pharmaceuticals in the environment. Their effectiveness however, depends on collaboration across several policy sectors and the adoption of the life cycle approach; taking action through pharmaceutical design, authorisation, manufacturing, prescription, over-the-counter purchases, consumer use (patients and farmers), collection and disposal, and wastewater treatment. A focus on preventive options early in a pharmaceutical's life cycle, may deliver the most long-term and large-scale environmental benefits.

1. Improve monitoring and reporting on the occurrence, fate, toxicity, and human health and ecological risks of pharmaceutical residues in order to lay the ground for pollution reduction policies. Consider the inclusion of environmental risks in the risk-benefit analysis of authorisation of new pharmaceuticals, and risk intervention and mitigation approaches for pharmaceuticals with high environmental risk.

2. Implement source-directed approaches, such as the sustainable design and procurement of pharmaceuticals, to prevent the release of pharmaceutical residues into water bodies.

3. Introduce use-orientated approaches, such as disease-prevention, improved diagnostics and restrictions on pharmaceuticals with high environmental risk, to reduce inappropriate and excessive consumption of pharmaceuticals.

4. Implement end-of-pipe measures, such as advanced wastewater treatment, public collection schemes for unused pharmaceuticals, and education campaigns, to safely dispose and remove pharmaceutical residues.

OECD Policy Recommendations on addressing pharmaceutical residues in freshwater

The OECD recommends government's take a collective, life cycle approach to managing pharmaceuticals in the environment. This means: i) designing and implementing a policy mix of source-directed, use-orientated and end-of pipe measures; ii) targeting stakeholders throughout the life cycle of pharmaceuticals; and iii) using a combination of voluntary, economic and regulatory instruments. A national pharmaceutical strategy and action plan to manage environmental risks should be developed in collaboration with relevant government departments, local authorities and other stakeholders, and be supported by a strategic financing strategy to ensure effective implementation.

Cross-cutting recommendations

- Identify potential environmental risks of existing and new active pharmaceutical ingredients (APIs) through intelligent and targeted monitoring and assessment strategies. Reduce unknowns on relationships between pharmaceuticals, and human and environmental health. The relative risk of APIs should also be compared with other pollutants (e.g. heavy metals, persistent organic pollutants and other contaminants of emerging concern) to achieve improvements in water quality and ecosystems in the most cost effective way.
- Encourage the uptake of new monitoring methods (see sections 2.4 and 2.5), modelling and decision-support tools to better understand and predict the risks, including mixtures as drivers of risk. Prioritise APIs and water bodies of highest concern.
- Increase access to data and information, and institutional coordination, to reduce knowledge gaps (see section 4.3.3).
- Adopt precautionary measures when scientific evidence is uncertain, and when the possible consequences of not acting are high.
- Factor in financing needs and measures to recover policy transactions costs, and factor in the capacity of government officials and stakeholders to implement policies.
- Educate and engage with the public to manage perceived and actual risks, and improve awareness and understanding.

Source-directed recommendations. Pharmaceutical life cycle stages: design, marketing authorisation, manufacturing, post-authorisation

- Develop clear and shared environmental criteria (and performance indicators) for sustainable 'green' procurement of pharmaceuticals.
- Consider expansion of the regulatory framework for good manufacturing practice to include mandatory environmental criteria.
- Develop drinking water safety plans, monitoring programmes of pharmaceuticals and incidence reporting to identify and prevent contamination and adapt policy to new science.
- Ensure Environmental Risk Assessment (ERA) robustness, consistency and transparency (see section 4.3.1). Establish a centralised database with independent regulatory oversight to share ERAs of APIs and prevent duplication efforts.
- Consider environmental risks in risk-benefit authorisation of human pharmaceuticals in order to manage and mitigate risks. Place more stringent conditions for putting a pharmaceutical on the market that is of high-risk to the environment (e.g. increased risk intervention and mitigation options, prescription only, eco-labelling, post-approval monitoring).
- Provide incentive structures to advance green and sustainable pharmacy (see section 4.3.2). A return on public investments in new pharmaceuticals should be considered when assessing support for the private sector in pharmaceutical development.
- Establish new business models for pharmaceuticals that balance access needs, appropriate use and adequate return. This is particularly important for new antibiotics and tackling antimicrobial resistance; current business models link profit (sales) with volume (consumption).

Use-orientated recommendations. Pharmaceutical life cycle stages: Prescription and use

- Reduce the incidence of infection and disease. Improved access to safe water supply, sanitation and hygiene is particularly important. Other important measures include improved stable and livestock handling, practitioner training, education campaigns and vaccinations.
- Reduce unnecessary use and release of pharmaceuticals. Improve diagnostics and delay prescription of pharmaceuticals when they are not immediately required. If not already in place, consider bans or restrictions on antibiotics for preventative use, and hormones as growth promoters, in the livestock and aquaculture sectors.
- Optimise the use of pharmaceuticals with effective diagnosis, dosing, personalised medicines and targeted delivery systems.
- Reduce self-prescription of pharmaceuticals with high environmental risk (e.g. antibiotics and pharmaceuticals that target the endocrine system) and illegal sales of pharmaceuticals.
- Promote best practices on the storage and use of livestock manure and slurry from livestock treated with pharmaceuticals.

End-of-pipe recommendations. Pharmaceutical life cycle stages: collection and disposal, and wastewater treatment and reuse

- End of pipe measures should only be used in complementary to source-directed and use-orientated measures. An over-emphasis on upgrading of wastewater treatment plant (WWTP) infrastructure is not a sustainable, optimal use of limited resources.

- Ensure value-for-money in investments in WWTP upgrades through evaluation and prioritisation, including achieving economies of scale (see section 3.5.1). Consider potential trade-offs (e.g. incomplete removal of APIs to varying degrees; generation of potentially toxic transformation products and sludge; increased energy, chemicals and carbon emissions).
- Factor in financing needs and cost-recovery mechanisms for capital and operation and maintenance costs of WWTP upgrades, including potential affordability issues with sanitation tariffs.
- Ensure appropriate collection and disposal of waste pharmaceuticals. Educate and engage with health professionals, veterinarians, consumers and farmers to raise awareness about inappropriate disposal of unused medications. Consider extended producer responsibility schemes to recover costs.
- Promote best practices on the use and disposal of biosolids (which may include toxic transformation products).

1 Defining the challenge of managing pharmaceuticals in water

This chapter characterises the diversity of pharmaceuticals, and their sources, mixtures and various entry pathways into the environment. It also summarises recent literature on the impacts of pharmaceuticals on water quality, human health and freshwater ecosystems, and makes the case for policy action.

1.1. Key messages

Pharmaceuticals are essential for human and animal health but have been recognised as an environmental concern when their residues enter freshwater systems. Due to demographic and epidemiological changes, the usage of pharmaceuticals has rapidly increased in OECD member countries. About 2 000 active pharmaceutical ingredients are being administered worldwide in prescription medicines, over-the-counter therapeutic drugs and veterinary drugs (Burns et al., 2018[1]).

The vast majority of them have not been evaluated for their occurrence, fate and possible impacts on water quality, human health and freshwater ecosystems. Municipal wastewater effluent is considered the most dominant pathway to freshwater bodies globally, however, emissions from manufacturing plants, hospitals, and intensive agriculture and aquaculture practices can be important sources locally.

As pharmaceuticals are designed to interact with living systems at low doses, even low environmental concentrations can be of concern. There is growing evidence of their occurrence in the environment and potential negative impacts. For example, steroid hormones in oral contraceptives have been proven in the laboratory to cause the feminisation of fish; psychiatric drugs, such as fluoxetine, can alter fish behaviour; and the use and discharge of antibiotics to water bodies is linked to antimicrobial resistance – a global health crisis. Some researchers stress the lack of human risk assessment regarding long-term and low-levels of pharmaceutical mixtures towards sensitive sub-populations (e.g. pregnant women, foetuses and children).

1.2. Introduction

Pharmaceuticals are synthetic or natural chemical compounds that are manufactured for use as prescription medicines, over-the-counter therapies, veterinary drugs and illicit drugs. Pharmaceuticals contain active ingredients that have been designed to have pharmacological effects and confer net benefits to society. The incidence of pharmaceuticals in the environment and the water cycle at trace levels (in the range of nanograms to low micrograms per litre) has been widely discussed and published in literature in the past decade. The increase in detection is largely attributable to advances in analytical techniques and instrumentation (WHO, 2012[2]) and the continuous increased use of pharmaceuticals.

About 2 000 active pharmaceutical ingredients (APIs) are being administered worldwide in prescription medicines, over-the-counter therapeutic drugs and veterinary drugs (Burns et al., 2018[1]). Their active ingredients comprise a variety of molecules produced by pharmaceutical companies in both the industrialised and the developing world at a rate of 100,000 tons per year (Weber et al., 2014[3]). The annual rate of increase in the development and approval of new APIs over the past five years has averaged about 43 in the United States (in 2018, the number approved was 59 – a record year) (Mullard, 2019[4]).

The continuous and increased production and use of pharmaceuticals has led to their widespread occurrence in the aquatic environment across the globe (Figure 1.1). Many APIs have been found worldwide in soils, biota, sediments, surface water, groundwater and drinking water. For example, research by Boxall and Wilkinson (forthcoming) tested 711 river sites in 72 countries for the presence of antibiotics and found antibiotics in 65% of them. In 111 of the sites, the concentrations of antibiotics exceeded safe levels, with the worst cases more than 300 times over the safe limit set by the AMR Industry Alliance.

Figure 1.1. Number of pharmaceuticals detected in surface water, groundwater or drinking water globally

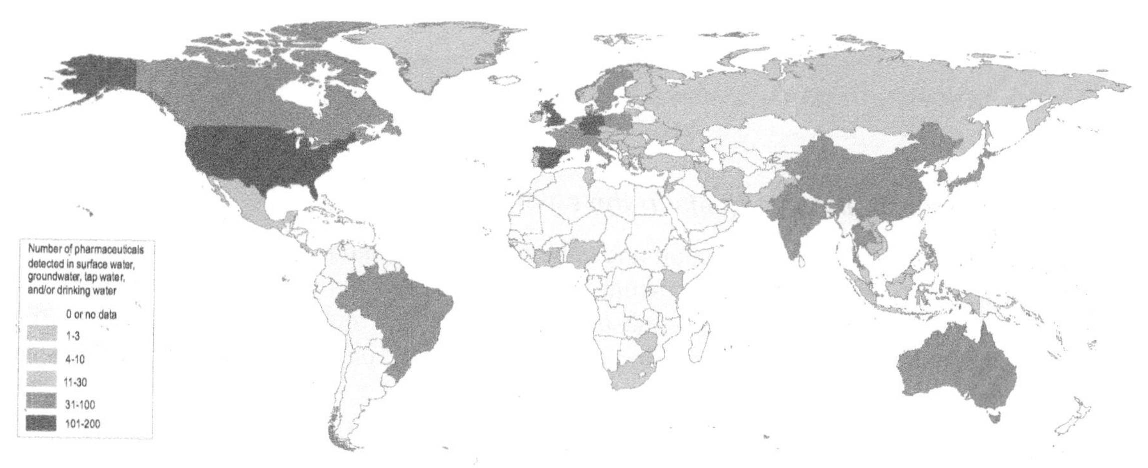

Source: (aus der Beek et al., 2016[5]).

As a consequence of APIs in the environment, there are recognised adverse effects on aquatic organisms, and recognised and undefined, long-term effects for humans through consumption of contaminated drinking water or food and antimicrobial resistance.

Pharmaceuticals in the environment are a challenge to manage for the following reasons:

- *Active pharmaceutical ingredients are designed to interact with a living system and produce a* pharmacological response at low doses, which makes them of environmental concern even at low concentrations. They are often designed to easily pass biological membranes and can interact with target molecules across a range of organisms. When exposed to non-target organisms in the environment, unintentional harmful impacts may occur. Furthermore, for APIs to be active at the target site, excess doses are regularly required to account for losses due to low uptake, availability issues and metabolisation of the API.

- Pharmaceuticals are designed to be stable (persistent or "pseudo-persistent") in order to reach and interact with target molecules (Khetan and Collins, 2007[6]). This means that either they are very slow to degrade or their constant use leads to continuous release into the environment at rates exceeding degradation rates (Bernhardt, Rosi and Gessner, 2017[7]). Hence, the chemical design elements for an effective pharmaceutical are contrary to what is desirable in the environment.

- Incentives are lacking for replacement of existing pharmaceuticals with "greener" alternatives. But in an era of growing human and ecosystem health threats from antimicrobial resistance and changes in ecosystem functioning, this neglect of environmental impacts of pharmaceuticals may be misplaced.

- Conventional wastewater treatment plants (WWTPs) are not designed to remove pharmaceuticals (although some APIs are removed by conventional wastewater treatment to a limited extent). Furthermore, veterinary pharmaceuticals used in agriculture and aquaculture can enter water bodies directly or via surface runoff (diffuse pollution).

- For most wildlife, exposure to pharmaceuticals in the environment could be long-term, potentially occurring via multiple exposure routes, and involving a mixture of substances.

- Managing the risks of pharmaceuticals in the environment requires a multi-disciplinary and multi-stakeholder approach.

In order to manage the risks of pharmaceuticals in the environment, it is first necessary to understand the origins, entry-pathways, sinks and concentration patterns of pharmaceuticals in the environment, and their effects on human and ecosystem health. The remainder of this chapter aims to do just that.

1.3. Origins, entry-pathways, sinks and concentration patterns of pharmaceuticals in the environment

1.3.1. A typology for pharmaceuticals in the environment

As mentioned in the previous section, human and veterinary pharmaceuticals, and their metabolites and transformation products, are ubiquitously present in water bodies. Pharmaceuticals are present in the environment as a consequence of pharmaceutical production and formulation, patient use, use in food production and improper disposal. After passing through the human or animal body, APIs are excreted either in an unchanged active form or as metabolites, which may be active or inactive, and have the potential for further breakdown into numerous transformation products in WWTPs or in the environment. Pharmaceuticals can disperse through the environment via multiple pathways as illustrated in Figure 1.2. The presence of pharmaceuticals in freshwater and terrestrial ecosystems can result in the uptake of pharmaceuticals into wildlife, and have the potential to bioaccumulate (Arnold et al., 2014[8]). Humans can subsequently be exposed through drinking water, and ingestion of pharmaceutical residues in plant crops, fish, dairy products and meat.

Figure 1.2. Major pathways of release of human and veterinary pharmaceuticals into the environment

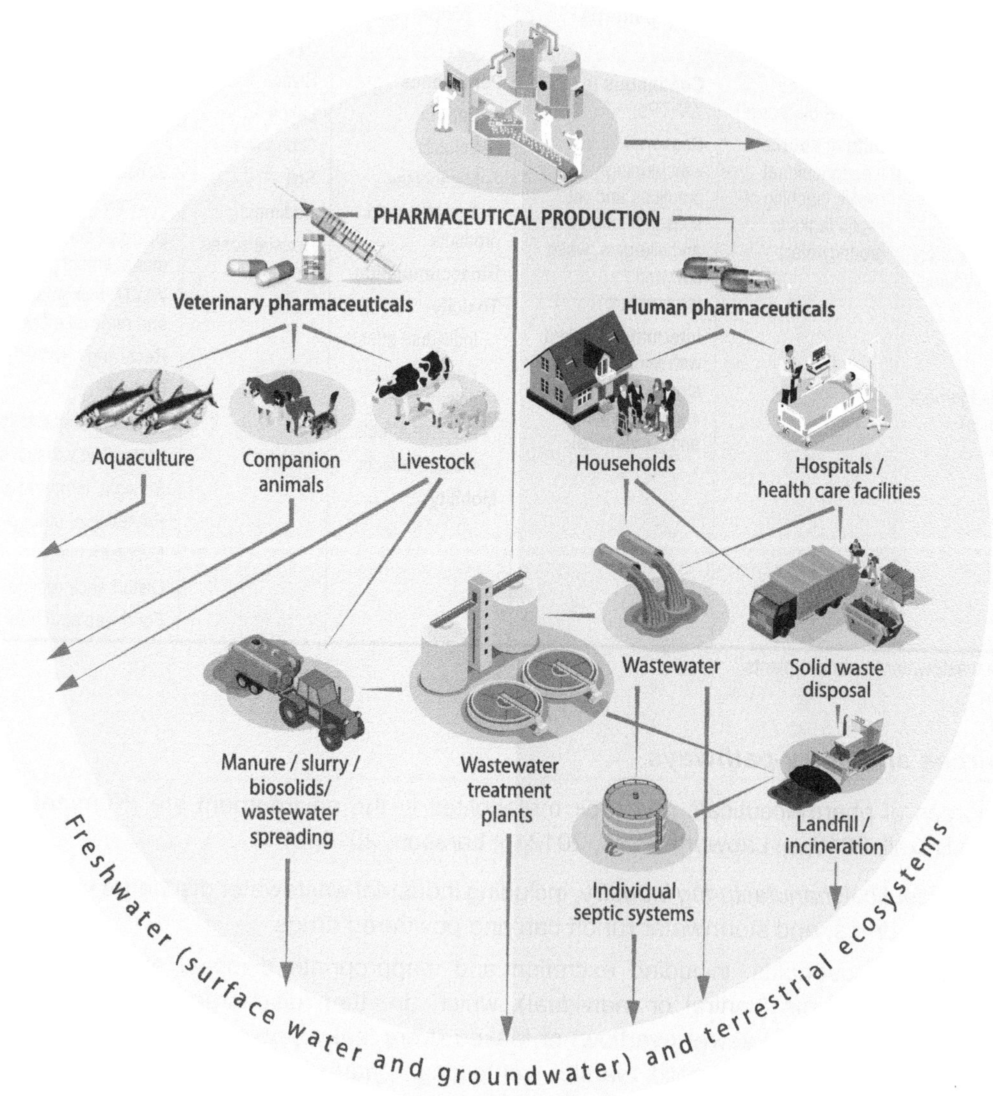

Note: Health care facilities include hospitals, clinics, pharmacies and elderly care homes.
Source: Author

The concentrations and impacts of pharmaceuticals in the environment depend on a combination of variables, including their use, and the toxicity, degradation, persistence and mobility properties of the pharmaceutical; source and timing of pollution; WWTP technology, operation and removal efficiency; agriculture and veterinary practices; sensitivity of the receiving environment and exposure history; and stochastic environmental conditions (Table 1.1).

Table 1.1. A typology for pharmaceuticals in the environment

Sources	Pathways	Concentration patterns	Pharmaceutical properties	Receiving environment type (sinks)	Concentration, context-dependent factors
Pharmaceutical manufacturing plants WWTPs - Municipal - Hospitals - Industry Agriculture (particularly intensive livestock farming) Aquaculture Septic tanks Waste management facilities (landfills)	**Point source** (WWTP discharge) **Diffuse source** (i.e. agricultural runoff, leaching of septic tanks to groundwater)	**Continuous** (e.g. WWTPs) **Seasonal** (linked with farming practices and with seasonal influenza and allergies, water flow and temperature) **Intermittent** (linked with rainfall events, stormwater overflow, irrigation patterns and pandemics)	**Persistence** - Half life - Solubility - Metabolites - Transformation products **Bioaccumulation** **Toxicity** - Individual effects - Population effects - Additive effects - Mixture effects **Mobility**	Rivers Lakes Groundwater Soil Sediment Coastal zones Oceans	Medical, agriculture and veterinary practices Illicit drug use Consumption rates Pharmaceutical properties Disposal and waste management practices WWTP technology, operation and removal efficiency Receiving environment type Climate Drainage characteristics Water flow variations Sunlight, temperature Presence of other pollutants Exposure history Disturbance regime Food web structure

Note: WWTPs: wastewater treatment plants.
Source: Author.

1.3.2. Sources and entry-pathways

The key sources of pharmaceuticals and their metabolites in the environment are (Kümmerer, 2009[9]; Monteiro and Boxall, 2010[10]; Lapworth et al., 2012[11]; Larsson, 2014[12]):

- *Pharmaceutical manufacturing industry*, including industrial wastewater discharge and solid wastes containing drugs, and stormwater runoff carrying powdered drugs.
- Consumers/households, including excretion and inappropriate disposal of pharmaceuticals to wastewater systems (central or individual), which are then discharged (treated or untreated, including as combined sewer overflow) or leaked (from leaky sewers and septic systems) to the environment. Household solid waste containing pharmaceuticals is another source of pharmaceuticals in the environment via landfills.
- Hospitals, including the discharge of wastewater and solid wastes.
- Agriculture and aquaculture, including: residual hormones and other drugs injected to poultry, cattle and fish; antibiotics added to livestock feed and waters; and runoff of livestock manure and slurry, and recycled wastewater and biosolids.

While the contribution of each emission source varies across regions and type of pharmaceuticals, it is generally accepted that, globally, the main route for human pharmaceuticals to the aquatic environment is via discharge of untreated or treated wastewater from households (Michael et al., 2013[13]; Weber et al., 2014[3]; Heberer and Feldmann, 2005[14]; Verlicchi, Al Aukidy and Zambello, 2012[15]; Verlicchi et al., 2010[16]). For veterinary pharmaceuticals, agriculture is the most significant source of water contamination via land application of livestock manure and slurry as irrigation water and fertiliser (Boxall, 2012[17]). For example, while practices differ across countries, globally more than 70% of all antimicrobials sold in 2013 were used in the agriculture sector, mainly as a growth promoter but also as a substitute for good hygiene (Van Boeckel, 2017[18]).

Pharmaceutical manufacturing

Pharmaceutical manufacturing facilities have been shown to release APIs into nearby streams and can be important pollution hotspots locally (Weber et al., 2014[3]; Larsson, de Pedro and Paxeus, 2007[19]) Extremely high pharmaceutical concentrations, in the order of mg/L have been detected in some industrial effluents and recipient streams, for example in India, China, USA, Korea and Israel (Larsson, 2014[12]). Environmental concentrations of pharmaceuticals discharged from manufacturing plants are generally much higher than excretion from humans (via WWTPs), and in some cases can greatly exceed toxic threshold concentrations. Although pollution from manufacturing is less widespread, discharges that promote the development of drug resistant microorganisms can have global consequences from a human health perspective (i.e. risk associated with antimicrobial resistance) (Larsson, 2014[12]).

Most API production takes place in emerging economies, mainly in Central and South America and in the Asia-Pacific region, which have become the global API production hubs. Consequently, this is where most of the pollution related to manufacturing occurs (BIO Intelligence Service, 2013[20]). Emissions from manufacturing facilities at production hubs leads to substantial discharges (in the order of several µg/L to mg/L) (Larsson, 2014[21]) causing contamination of surface water, sediment (Kristiansson et al., 2011[22]), groundwater and drinking water wells (Fick et al., 2009[23]). For example, the effluent of one WWTP serving 90 manufacturers of bulk drugs in Patancheru, Hyderabad, India was found to have levels of ciprofloxacin as high as 32 mg/L (Larsson, de Pedro and Paxeus, 2007[24]), which is considerably higher than levels found in the blood of patients taking antibiotics. The most striking finding from this Indian WWTP was that all bacteria isolated inside the facility were found to be multi-resistant to antibiotics which can have global consequences for the spread of antimicrobial resistance (Johnning et al., 2013[25]; Marathe et al., 2013[26]) (for introduction to antimicrobial resistance see Box 1.7). Furthermore, in the receiving lake and up to 17 km downstream from the WWTP, the levels of resistant genes were high (Kristiansson et al., 2011[22]).

Within developed nations, emissions of pharmaceuticals from manufacturing facilities (including facilities that formulate finished pharmaceutical products from APIs imported in emerging economies) are less widespread compared to emissions from WWTPs. However, pollution downstream of manufacturers has been observed at EU monitored sites (e.g. the Rhine, Lake Leman) (BIO Intelligence Service, 2013[20]) and when investigating effluents in the U.S (Scott et al., 2018[27]) and other OECD countries (Larsson, 2014[21]).

Household and hospital consumption (usage)

The consumption stage of the life cycle of pharmaceuticals is considered the greatest contributor to the environmental load of pharmaceutical residues in water in OECD countries. Pharmaceuticals administrated to humans or animals are excreted via urine and faeces, with an estimated 30 to 90% of oral doses generally excreted as active substances (BIO Intelligence Service, 2013[20]). However, the nature and amount of medicinal residues mainly depend on the volumes and nature of the administered substances, their modes of administration, and metabolisation rates. The usage of pharmaceuticals varies across regions in terms of commonly used substances, clinical practice and prescription patterns.

The dominant source of human pharmaceuticals in the environment is from households; pharmaceutical use in hospitals and nursing homes is estimated to account for a few percent of the total release from a city (Azuma et al., 2016[28]; Lacorte et al., 2018[29])). However, this varies widely depending on the type of pharmaceutical. Some substances are only intended for use in hospitals, whilst others are taken, or excreted, at home. For instance, hospitals are the dominant (70-90%) source of anti-cancer drugs, endocrine therapy and contrast media (although they may be excreted at home after discharge from hospital), while households are the dominant source of painkillers, blood pressure medicine (BIO Intelligence Service, 2013[20]) and anti-inflammatories (Daughton and Ruhoy, 2009[30]). Knowledge gaps exist regarding the sales and consumption of over-the-counter and self-prescribed pharmaceuticals.

Municipal WWTPs collect and concentrate a variety of human pharmaceuticals (and their metabolites) administered in households, hospitals and elderly care homes. Conventional WWTPs are not designed to remove pharmaceuticals or their metabolites (Box 1.1); WWTPs are primarily designed to remove pathogens, suspended solids and organic and inorganic matter, rather than the removal of the increasing numbers of modern chemicals, including pharmaceuticals at low concentrations (Melvin and Leusch, 2016[31]). WWTPs can therefore release APIs and metabolites to the environment, depending on the level and type of wastewater treatment, (Yang et al., 2017[32]) in unchanged forms or as transformation products. Unused medicines that are improperly disposed of in sinks and toilets also end up in municipal wastewater.

Box 1.1.Limitations of wastewater treatment plants for removal of pharmaceuticals

The degree of removal of different pharmaceuticals in wastewater treatment plants (WWTPs) is highly variable depending on the type of pharmaceuticals entering the system, their physico-chemical properties, and the removal efficiency of WWTP technology. There are large discrepancies in removal efficiencies of pharmaceuticals in WWTPs between countries, and even between WWTPs within the same country (Tran, Reinhard and Gin, 2018[33]). No single technique has been found to remove all relevant pollutants from wastewater (Hollender et al., 2009[34]; Melvin and Leusch, 2016[31]; Behera et al., 2011[35]) (Verlicchi, Al Aukidy and Zambello, 2012[15]).

In a review by Deblonde et al. (2011[36]), the removal rates of pharmaceuticals following wastewater treatment ranged from 0% (contrast media) to 97% (psychostimulant). The removal rate for antibiotics was about 50%. Analgesics, anti-inflammatories and beta-blockers were some of the most resistant to treatment (30–40% removal rate).

In the UK, as part of the Chemical Investigation Programme, the concentrations of 19 APIs and 4 metabolites were monitored 20 times in the influent and effluent of 45 WWTPs over a two year period (2015-2017). The results, published by Comber et al. (2018[37]), show that the majority of substances studied were removed to a high degree, although with significant variation, both within and between WWTPs. Poorer removal (between influent and effluent) was observed for ethinyloestradiol, diclofenac, propranolol, the macrolide antibiotics, fluoxetine, tamoxifen and carbamazepine. All except the last two of these substances were present in effluents at concentrations higher than their respective estimated PNEC. Based on available dilution data, as many as 890 WWTPs in the UK (approximately 13% of all WWTPs) may cause exceedances of estimated riverine PNECs after mixing of their effluents with receiving waters. If the estimated PNECs are a guide to regulatory limits, then there is potential for localised non-compliance in surface waters, at least in the case of ethinyloestradiol, diclofenac, ibuprofen, propranolol and the macrolide antibiotics (Comber et al., 2018[37])

Sources: (Comber et al., 2018[37]) (Deblonde, Cossu-Leguille and Hartemann, 2011[36]) (Hollender et al., 2009[34]; Melvin and Leusch, 2016[31]; Behera et al., 2011[35]) (Verlicchi, Al Aukidy and Zambello, 2012[15]) (Gardner et al., 2012[38]).

Pharmaceuticals improperly disposed of in the household garbage end up in landfills which can eventually be transferred to surface or groundwater bodies if there is no collection of landfill leachate (Tong, Peake and Braund, 2011[39]; Saad et al., 2017[40]; Barnes et al., 2004[41]). A large number of the prescription items dispensed every year are not taken or administered, and become waste. For example, in the US, it is estimated that about one-third of the four billion prescription items annually become waste (Product Stewardship Council, 2018[42]). Over-prescription, self-medication (over-the-counter pharmaceuticals) and misdiagnosis of symptoms can increase the amount of APIs administered and improperly disposed of. There is little information about the contribution of improper disposal of pharmaceuticals in relation to other sources of pollution (BIO Intelligence Service, 2013[20]).

Agriculture and aquaculture

Although veterinary pharmaceuticals may benefit the health and welfare of domestic animals and the efficiency of intensive food animal and fish production, they can contaminate the water resources through manufacturing, treatment of animals, and disposal of carcasses, offal, effluent, manure and unused products (Boxall, 2010[43]). Veterinary pharmaceuticals used in aquaculture directly enter surface waters (Weber et al., 2014[3]).

A wide range of veterinary pharmaceuticals are used in the agriculture sector, for example, antimicrobial medicines (antibiotics, antiprotozoals and parasiticides) and hormones. The overuse of pharmaceuticals in industrial farming results in a significant release of their residues into soil, groundwater and surface water. The main entry pathways are from the use of animal manure as a fertiliser, animal waste storage and disposal tanks (FAO, 2018[44]). Moreover, the reuse of biosolids as fertiliser or irrigation of recycled wastewater from WWTPs onto agricultural land will eventually spread human pharmaceuticals to the surrounding environment (FAO, 2018[44]). Over time, residues from these drugs accumulate in the soil or drain into surface and groundwater; where they may also be taken up by plants (Weber et al., 2014[3]; Carter et al., 2014[45]).

The release of oestrogen hormones from livestock is the most significant source of oestrogens to the environment. The discharges of oestrogen from livestock are estimated at 83,000 kg per year in the U.S and EU alone (Adeel et al., 2017[46]). This compares to the annual global discharge of oestrogens from oral contraceptives, estimated to be approximately 30,000 kg of natural steroidal oestrogens (E1,E2,E3) and 700 kg of synthetic oestrogens (EE2) (calculations based on 7 billion people). Oral administration of the synthetic steroid hormone17-α-methyl-testosterone in fish hatcheries to produce mono-sex of certain fish species is commonly practiced in south-east Asia, with potential release of effluents to the surrounding water (Rico et al., 2012[47]).

Knowledge gaps exist regarding the sales and consumption of veterinary pharmaceuticals, thus it is difficult to estimate the total amount used and released to the environment. It is however, recorded that more than 70% of the total volume of all medically-important antibiotics in the United States[1] (and over 50% in most countries globally) are sold for use for livestock (Review on Antimicrobial Resistance, 2015[48]). One important reason for the difference between human and livestock antibiotic use is that human use is commonly for *treatment* of infections, whereas livestock use is commonly for disease *prevention* and to marginally improve growth rates (Martin, Thottathil and Newman, 2015[49]).

Antibiotics are also used in aquaculture to improve the health status of the cultured organisms, to prevent or treat disease outbreaks, and as a growth promoter. Aquaculture systems are hydrologically connected with the surrounding water, thus a considerable amount of antibiotics (70-80%) may be released to the surrounding water (Review on Antimicrobial Resistance, 2015[48]). Almost 90% of the global aquaculture production takes place in Asia, primarily in tropical and subtropical regions (FAO, 2016[50]), where the use of 36 antibiotics has been documented (Rico et al., 2012[47]).

1.3.3. Concentration patterns

Concentration patterns of pharmaceuticals in the environment can be classified as: i) *continuous* (e.g. from WWTPs), ii) *seasonal* (linked with farming practices, seasonal influenza and allergies, flow rates and temperature), iii) and *intermittent* (linked with rainfall events, stormwater overflow and irrigation patterns).

Effluent from WWTPs is continuously discharged into surface water, with contaminant loads varying due to the number of households and hospitals connected, the level of wastewater treatment, disease outbreaks and stochastic environmental conditions. Data on the occurrence and concentrations of some pharmaceuticals in effluents from WWTPs and in surface waters show that pharmaceutical concentrations fluctuate widely (Table 1.2), most probably due to different pharmaceutical doses applied in various regions and inconsistent efficiency of wastewater treatment (Pal et al., 2010[51]). A range of antibiotics, analgesics, anti-inflammatories, anticonvulsants, beta-blockers and blood lipid modifying agents have been detected in various concentrations in both WWTP effluent and receiving surface waters in North America, Europe, Australia and Asia. Antibiotics are of particular concern with levels in effluent in each of these regions greater than the predicted no effect concentration (PNEC), and high proportions of the parent pharmaceutical compound detected following wastewater treatment (Table 1.2) (Pal et al., 2010[51]). Other reviews report similar results, with the exception of trimethoprim and ibuprofen reported at one higher level of magnitude in wastewater effluent in North America (e.g. (Tran, Reinhard and Gin, 2018[33])).

Table 1.2. Data on the occurrence and concentration levels of various pharmaceuticals from wastewater treatment plant effluents, and in freshwater rivers and canals

Pharmaceutical compound / molecules	Range in concentration (ng/L)						Lowest PNEC (ng/L)	% of parent compound excreted
	North America		Europe		Asia & Australia[1]			
	WWTP effluent	Surface water	WWTP effluent	Surface water	WWTP effluent	Surface water		
Antibiotics								
trimethoprim	<0.5–**7900**	2 – 212	99 -**1264**	0 - 78.3	58 – 321	4 – 150	1000	≥70
ciprofloxacin	**110 - 1100**	-	40 - 3353	-	**42 – 720** [2]	ND – **1300** [3]	20	≥70
sulfamethoxazole	5 - 2800	7 - 211	91 - 794	<0.5 - 4	3.8 - 1400	1.7 - 2000	20,000	6-39
Analgesics and anti-inflammatories								
naproxen	<1 – 5100	0 – 135.2	450 – 1840	<0.3 – 146	128 – 548	11 – 181	37,000	-
ibuprofen	220 - 3600	0 - 34	134 - **7100**	14 - 44	65 - 1758	28 - 360	5000	≤5
ketoprofen	12 - 110	-	225 - 954	<0.5 - 14	-	<0.4 – 79.6	15.6 x 10^6	-
diclofenac	<0.5 – 177.1	11 - 82	460 - 3300	21- 41	8.8 - 127	1.1 – 6.8	10,000	6 - 39
acetylsalicylic acid	47.2 - 180	70 – 121	40 – 190	<0.3 – 302	9 – 2098	-	-	6 - 39
mefanemic acid	-	-	1 - 554	<0.3 - 169	4.45 - 396	<0.1 – 65.1	-	-
acetaminophen	-	24.7 – 65.2	59 - 220	<12 - 777	1.8 - 19	4.1 - 73	9200	≤5
Anticonvulsants								
carbamazepine	111.2 - 187	2.7 – 113.7	130 - 290	9 - 157	152 - 226	25 – 34.7	25,000	≤5
Beta-blockers								
propranolol	-	-	30-44	20	50	-	500	<0.5
atenolol	879	-	1720	314	-	-	10 x 10^6	50 - 9
Blood lipid modifying agents								
clofibrate acid	ND – 33	3.2 – 26.7	27 – 120	1 – 14	154	22 – 248	12,000	-
gemfibrozil	9 - 300	5.4 - 16	2 – 28,571	-	3.9 - 17	1.8 – 9.1	100,000	-
Bezafibrate	ND - 260	-	233 - 340	16 - 363	-	-	100,000	40 - 69

Notes: PNEC: predicted no effect concentration; Bold denotes measured concentrations greater than PNEC (note that effluent levels are not compared to PNEC); ND: not detected; Dashed line: not reported.

1. Pharmaceutical concentrations for Australia are based on a single study from one region (Queensland) (Watkinson et al., 2009[52]) and this may not be representative of levels in other parts of Australia.

2. There is no data on ciprofloxacin in WWTP effluent in the Australian study.

3. Concentrations found in surface water in the Australian study ranged from below detection to below 1300 ng/L.

Source: (Pal et al., 2010[51]).

The majority of studies report peaks of human pharmaceutical concentrations during cold seasons (Lindholm-Lehto et al., 2016[53]) (Singer et al., 2014[54]) (Sun et al., 2014[55]) (Yu, Wu and Chang, 2013[56]) (Kot-Wasik, Jakimska and Śliwka-Kaszyńska, 2016[57]). Seasonal peaks of anti-inflammatories, analgesics and antibiotics in WWTP influent and effluent during cold seasons can be explained by increased usage, but also by reduced WWTP removal capacity (reduced microbial activity of activated sludge) due to cooler temperatures (Sun et al., 2014[55]).

Seasonal or intermittent peaks from WWTPs can also be observed during pandemics or other events when the use of a certain set of pharmaceuticals increase within a short timespan. For instance, (Singer et al., 2014[54]) showed that antivirals, antibiotics and decongestants in WWTP effluents and the River Thames, UK increased during November 2009 - the autumnal peak of the 2009 influenza pandemic. Increased effluent concentrations of antihistamines during the spring can be attributed to the onset of spring allergies (Vatovec et al., 2016[58]). Increased pharmaceutical concentrations may also be observed during high rainfall events when sewage bypasses treatment through combined sewer overflows.

Several studies report increased concentrations of pharmaceuticals in water bodies as a result of reduced river flows. For instance, human pharmaceuticals significantly increased during dry weather conditions in South Wales, UK due to reduced dilution with surface water flows (Kasprzyk-Hordern, Dinsdale and Guwy, 2008[59]). Temporal peak concentrations of pharmaceuticals in surface water, as a result of lower rainfall and therefore reduced dilution of WWTP discharges, have also been seen to affect the concentration patterns of pharmaceuticals in drinking water (Padhye et al., 2014[60]).

Concentration peaks of veterinary pharmaceuticals in water bodies, with origins from soil and agricultural practices, are primarily driven by rainfall and the properties of the receiving water body. Most studies suggest that the soil serves as a reservoir and peak concentrations of pharmaceuticals in water are, in general, associated with overland runoff caused by rain events and soil erosion (Jaimes-Correa, Snow and Bartelt-Hunt, 2015[61]) (Bernot, Smith and Frey, 2013[62]) (Forrest et al., 2011[63]) (Lissemore et al., 2006[64]). In addition, variations in the use of veterinary pharmaceuticals may be driven by agricultural practices and increased usage during certain periods (such as calving or lambing), and timing of manure, slurry and irrigation applications. Pharmaceuticals released during summer are expected to undergo more rapid degradation in the environment due to increased temperature and more intense sunlight (Lindholm-Lehto et al., 2016[53]).

1.3.4. Environmental sinks

The occurrence of certain pharmaceuticals in the environment has been acknowledged for several decades. In the environment, active pharmaceutical ingredients are found in surface waters, groundwater, soil, manure, biota, sediment, drinking water and the food chain (Benotti et al., 2009[65]; Michael et al., 2013[13]; Daughton and Ternes, 1999[66]; de Jongh et al., 2012[67]; Mompelat, Le Bot and Thomas, 2009[68]; Monteiro and Boxall, 2010[10]; Verlicchi, Al Aukidy and Zambello, 2012[15]; Lapworth et al., 2012[11]). Thanks to improved analytical techniques and established laboratories, low-levels of environmental pollutants have increasingly been detected in all regions of the world (Table 1.3). In a global review by aus der Beek et al. (2016[5]), a total of 631 human and veterinary pharmaceuticals (including 127 transformation products) were detected in surface, ground and drinking waters in 71 countries. Table 1.3 represents frequently analysed pharmaceuticals (antibiotics, analgesics, oestrogens and blood lipid modifying agents). It does not provide the full picture of pharmaceutical occurrence since other pharmaceuticals have relatively limited monitoring.

Concentrations of pharmaceuticals in surface freshwater are the most documented, and represent the environmental sink receiving most discharges from WWTP effluent, agriculture and aquaculture. Surface waters, in general, contain higher levels and a greater range of pharmaceuticals in comparison to groundwater bodies (Focazio et al., 2008[69]) (Vulliet and Cren-Olivé, 2011[70]). Countries that use surface water as a source of drinking water tend to have higher concentrations of pharmaceuticals in drinking water in comparison to those using groundwater as a drinking water source (BIO Intelligence Service, 2013[20]). However, groundwater is also reported as an important sink, and under certain conditions, may pose a prolonged threat to drinking water sources due to long groundwater residence times (Lapworth et al., 2012[11]). For instance, a study by Barnes et al. (2008[71]) identified pharmaceuticals and personal care products in 81% of sampled groundwater sites in 18 states of the U.S.

The marine environment is less characterised for its occurrence of pharmaceuticals compared to freshwater (Arpin-Pont et al., 2016[72]) but there is increasing evidence that pharmaceuticals are present in, and have the potential to impact, marine and coastal environments (Gaw, Thomas and Hutchinson, 2014[73]; Fabbri and Franzellitti, 2016[74]). For example, estuarine systems receiving chronic inputs of trace concentrations of the antimicrobial tylosin, as well as other antibiotics, may experience reductions in benthic microalgal biomass and primary productivity (Pinckney et al., 2013[75]). Examples of laboratory reported adverse effects for analgesics on marine organisms include: reduced feeding rates (Solé et al., 2010[76]), impacts on survival (Guler and Ford, 2010[77]), reduced mussel byssus strength (Ericson, Thorsén and Kumblad, 2010[78]), oxidative and neurotoxic effects (Mezzelani et al., 2016[79]), and changes in immune response (Solé et al., 2010[76]; Mezzelani et al., 2016[79]) and biochemical markers (Gonzalez-Rey and Bebianno, 2014[80]).

Monitoring is still lacking in both coverage and frequency in developing economies (Puckowski et al., 2016[81]; Madikizela, Tavengwa and Chimuka, 2017[82]). Where studies have been undertaken, higher concentrations of pharmaceutical pollutants have been found (in comparison to developed nations), which may reflect a lack of wastewater treatment infrastructure (Segura et al., 2015[83]).

Table 1.3. Pharmaceuticals found in surface, ground and drinking waters in all regions of the world

Global average and maximum concentration for surface water

Pharmaceutical compound	Theraputic group	Number of countries[1]	Average (μg/L)[2]	Maximum (μg/L) [2]
diclofenac	Analgesic	50	0.032	18.74
carbamazepine	Anticonvulsant	48	0.187	8.05
ibuprofen	Analgesic	47	0.108	303.0
sulfamethoxazole	Antibiotics	47	0.095	29.0
naproxen	Analgesic	45	0.050	32.0
oestrone (E1)	Oestrogens	35	0.016	5.0
oestradiol (E2)	Oestrogens	34	0.003	0.012
ethinylestradiol (EE2)	Oestrogens	31	0.043	5.9
trimethoprim	Antibiotics	29	0.037	13.6
paracetamol	Analgesic	29	0.161	230.0
clofibric acid	Blood lipid modifying agents	23	0.022	7.91
ciprofloxacin	Antibiotics	20	18.99	6500
ofloxacin	Antibiotics	16	0.278	17.7
oestriol	Oestrogens	15	0.009	0.48
norfloxacin	Antibiotics	15	3.457	520.0
acetylsalicylic acid	Analgesic	15	0.922	20.96

1. Number of countries: countries worldwide with positive detection of pharmaceuticals in surface water, groundwater or drinking water.
2. Average and maximum concentrations are of measured surface water concentrations.
Source: (aus der Beek et al., 2016[5]).

1.3.5. Future projections on pharmaceuticals in the environment

In OECD countries, pharmaceutical consumption has rapidly grown over the last decade, owing to aging populations, epidemiological changes (increasing need, ability and expectation to treat ageing-related and chronic diseases), and changes in clinical practice (recommendations of earlier treatment, higher dosages or prolonged treatment) (Belloni, Morgan and Paris, 2016[84]). This trend is expected to continue and is reflected in a projected growth rate of the pharmaceutical industry of 6.5% per year by 2022 (UN Environment, 2019[85]). As an example, the growth prognosis for consumption of prescription pharmaceuticals in Germany is presented in Box 1.2. Urbanisation is another factor compounding risks of pharmaceuticals in the environment; a larger group of ageing people will be discharging via a single wastewater treatment plant.

Box 1.2. Pharmaceutical usage in the context of demographic change, Germany

The German Association of Energy and Water Industries (BDEW) investigated the future trends of human pharmaceutical usage in Germany. The prediction was based on a prognostic model of human pharmaceutical usage based on a population projection, differentiated by age and gender-specific consumption. The study concludes that as a result of the demographic change and an increased consumption per capita, pharmaceutical usage is projected to increase by 43-67 percent by the year 2045 (from a baseline of 2015) (Figure 1.3).

Pharmaceutical usage is dominated by older population groups (age 80-85), consuming about 20 times more compared to those of younger age (age 20-25). An ageing population is thought to be the main driver; of the total population, people over the age of 60 years will increase by 9% by 2045. An increase in consumption of pharmaceuticals is also expected in younger age groups (Civity, 2017[86]).

Figure 1.3. Growth prognosis for consumption of prescription pharmaceuticals for human use in Germany

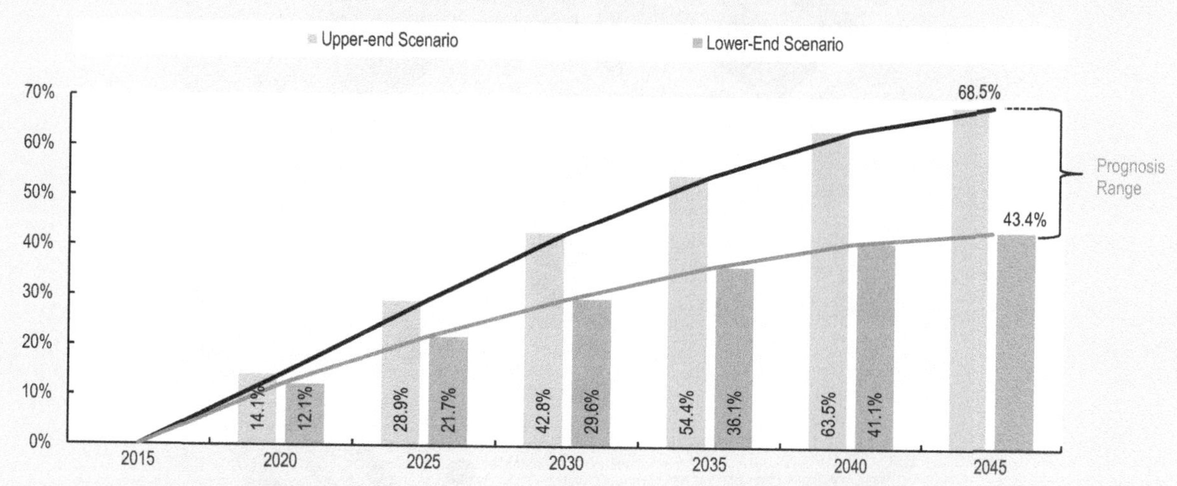

Source: Pharmaceutical usage in the context of demographic change: A study by Civity Management Consultants commissioned by BDEW (Civity, 2017[86]).

As global meat and fish production is projected to increase[2] (Alexandratos and Bruinsma, 2012[87]; OECD/FAO, 2018[88]), as well as demand for companion animal pharmaceuticals (NOAH, 2018[89]), so will the use of veterinary pharmaceuticals increase. For example, antibiotics administered to livestock animals in feed are projected to increase by 67% worldwide by 2030 (from 2015 levels) (Van Boeckel et al., 2015[90]). Much of this increase will come in emerging economies.

In addition, changes in climate are likely to affect the amounts and types of pharmaceuticals used and released to water bodies. Substantially higher pharmaceutical use appears inevitable as climate change stresses native life, thereby enabling pathogens to cause disease (Cavicchioli et al., 2019[91]). For example, non-communicable diseases (e.g. cardiovascular disease and mental illness) and respiratory, water-borne, vector-borne and food-borne toxicants and infections are expected to become more common (Cavicchioli et al., 2019[91]) (Redshaw et al., 2013[92]) (Figure 1.4). Climate change is predicted to increase the rate of antibiotic resistance of some human pathogens (MacFadden et al., 2018[93]).

Climate-related environmental changes have already been associated with a rise in the incidence of chronic diseases in the Northern Hemisphere (Redshaw et al., 2013[92]). Certain vector-borne diseases, such as bluetongue, an economically important viral disease of livestock, have already emerged in Europe in response to climate change, and larger, more frequent outbreaks are predicted to occur in the future (Jones et al., 2019[94]). Millions of people are predicted to be newly at risk under climate change to mosquito-borne and tick-borne diseases (Cavicchioli et al., 2019[91]).

Figure 1.4. Climatic conditions causing non-communicable illnesses and associated pharmaceutical treatments

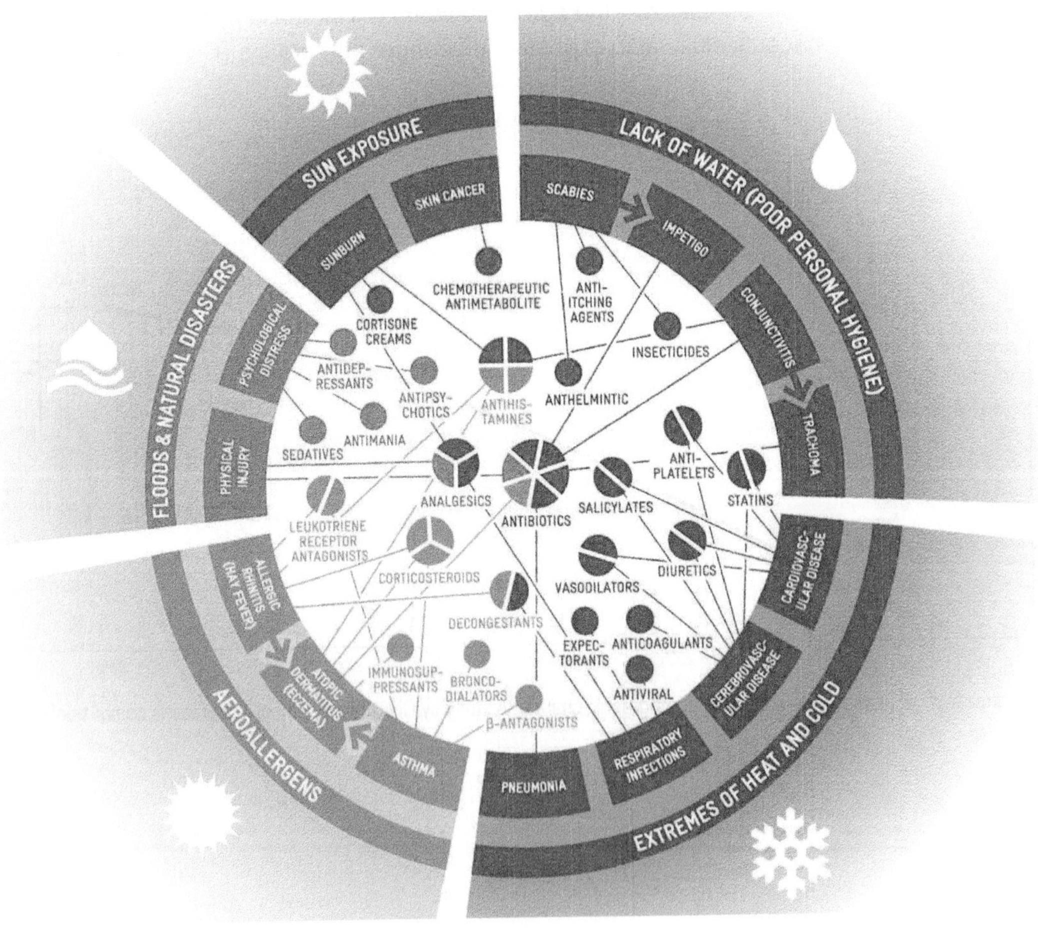

Source: (Redshaw et al., 2013[92]).

Climate change will also entail changes in farming practice. For example, increasing extreme weather events will mean livestock populations in some regions will be subjected to thermal stress and waterlogged pastures. This may in turn, lead to increased indoor housing of animals, facilitate the introduction of new pathogens, vectors, and hosts, and thus lead to increased use of veterinary pharmaceuticals (Boxall et al., 2009[95]).

Overall, it is anticipated that climate change will affect the fate and transport of pharmaceuticals, and result in an increase in the likelihood of pharmaceuticals entering the environment. Increases in temperature and changes in moisture content are likely to reduce the persistence of pharmaceuticals (i.e. increase degradation rates) but changes in hydrological characteristics are likely to increase the potential for contaminants to be transported to water bodies. For example, extreme weather events will mobilise contaminants from soils and manure in agriculture systems, potentially increasing their bioavailability, and heavy rainfall events will trigger stormwater overflows, bypassing WWTPs (Boxall, 2012[17]). Conversely, in areas affected by lower rainfall and increased water scarcity, the dilution potential of pharmaceutical residues and metabolites in surface water will be reduced.

As a result of the above projected trends, unless adequate measures are taken to manage the related risks, pharmaceuticals will increasingly be released into the environment (Weber et al., 2014[3]).

1.4. Effects of pharmaceuticals in the environment on human and freshwater ecosystem health

1.4.1. Proven and potential adverse effects and costs

It is challenging to assess the potential long-term risks of trace amounts of pharmaceuticals in the environment, especially given that water resources (including drinking water) are often not systematically monitored for pharmaceutical residues, and their cause and effect. While our understanding of the environmental impact for several pharmaceuticals is increasing[3] (Donnachie, Johnson and Sumpter, 2016[96]), the vast majority of pharmaceuticals have not been evaluated for their long-term toxicity, occurrence or fate in the environment, and it is therefore difficult to generalise the risk they may give rise to. Furthermore, whilst pharmaceuticals are stringently regulated for efficacy and patient safety, the adverse effects they may have in the natural environment have not yet been sufficiently studied and are not covered by an international agreement or arrangement (Weber et al., 2014[3]).

Little evidence of risk to human health exists. Currently, quantitative risk assessment studies indicate no appreciable human health risks associated with exposure to pharmaceuticals in drinking water (WHO, 2012[2]). Concentrations of pharmaceuticals found in drinking water are generally greater than 1000-fold below the minimum therapeutic dose, which is the lowest clinically active dosage. Likewise, other environmental risk assessments of human exposure of pharmaceuticals suggests no appreciable risk to human health e.g. (Silva et al., 2017[97]) (Bercu et al., 2008[98]) (Cunningham, Binks and Olson, 2009[99]) (Johnson et al., 2008[100]) (Kostich, Batt and Lazorchak, 2014[101]) (Baken et al., 2018[102]) (Houtman et al., 2014[103]). However, it is important to note that uncertainties and particular concerns still exist. The minimum therapeutic dose for a certain beneficial effect of a pharmaceutical to occur is not related to the effect concentration of other, unintended effects. Some cytotoxic drugs used in anti-cancer treatment will not have a safe lower level if they interact directly with DNA. Many researchers stress the lack of knowledge regarding long-term and low-level exposure of pharmaceutical mixtures in the environment. In most studies, the targeted population consists of healthy adults. But for sensitive populations, such as children, pregnant women, foetuses, and people with allergies and chronic diseases, the risk could be greater (Collier, 2007[104]; Snyder et al., 2008[105]; Johnson et al., 2008[106]; BIO Intelligence Service, 2013[107]). Furthermore, most studies also assume the risk posed by a single compound is comparable to one posed by a mixture (Kümmerer, 2009[9]). Little evidence exists on the potential human health risks from consumption of food containing APIs, although there is evidence proving the potential for bioaccumulation and transfer of APIs through the food web (see section 1.4.2).

Nevertheless, the presence of pharmaceuticals in the environment has raised concerns among drinking water regulators, governments, water suppliers and the public, regarding the proven and potential risks to human and environmental health. Certain pharmaceuticals have been proven to cause undesired adverse effects on ecosystems, including mortality, and changes to physiology, behaviour, reproduction. For example, diclofenac in the environment has resulted in the endangerment of species of vultures, antidepressants have been shown to alter fish behaviour (Box 1.3) and endocrine disrupting pharmaceuticals can interfere with fish reproduction (Box 1.4) (BIO Intelligence Service, 2013[107]; Santos et al., 2010[108]; Oaks et al., 2004[109]; Green et al., 2004[110]). The German Environment Agency (UBA) estimate that 10% of pharmaceutical products indicate a potential environmental risk (Küster and Adler, 2014[111]). Of greatest concern are hormones, antibiotics, analgesics, antidepressants and anticancer pharmaceuticals used for human health, and hormones, antibiotics and parasiticides used as veterinary pharmaceuticals (Küster and Adler, 2014[111]).

In a study by Gunnarsson et al. (2019[112]),, environmental risks of a range of pharmaceuticals with a full set of ecotoxicity data (excluding antibiotics) were assessed based on a combined analysis of drug toxicity and predicted environmental concentrations based on European patient consumption data. Pharmaceuticals that target the endocrine system represented the highest potency and greatest environmental risk (PEC/PNEC >10). Propranolol (beta blocker) and fluoxetine (antidepressant) were found to be of moderate environmental risk (PEC/PNEC >1). Most drugs (> 80%) with a full set of ecotoxicity data indicated a low environmental risks for the endpoints assessed in a European context.

A summary of potential adverse effects of some pharmaceuticals in the environment on freshwater ecosystem and human health is presented Table 1.4.

Box 1.3. Behaviour alternations in fish from exposure to antidepressants

A growing number of studies link exposure to antidepressants to behavioural alterations in fish, including changes in feeding rate, mating success, parental care, predator avoidance, aggression, reduced social interaction, foraging efficiency, dispersal and migration. These disrupted ecological interactions may affect the food web structure and functions of the ecosystem (Brodin et al., 2014[113]). For instance, Kellner et al. (2016[114]) showed in laboratory experiments that three-spine stickleback fish exposed to the antidepressant citalopram increased their locomotor activity and decreased bottom dwelling. The fish became bolder and less sensitive to stress, which made them more vulnerable to predators.

Sources: (Brodin et al., 2014[113]); (Kellner et al., 2016[114]).

Box 1.4. An introduction to endocrine disrupting chemicals and their potential effects on human and ecosystem health

Natural hormones such as oestrogen and progestin are essential for the function of the female reproductive system, while androgens regulate the male sex organs (WHO, 2012[115]). There are natural and synthetic hormones typically found in contraceptives, hormonal therapies and veterinary medicine (as growth-regulators of farmed animals) which are intended to interfere with the natural hormonal system of the body, but when released to the aquatic environment they pose a threat to organisms and their reproduction systems. Other pharmaceuticals that have showed potential to act as endocrine active substances include triclosan (antibacterial and antifungal), fluoxetine (antidepressants) and diclofenac (anti-inflammatory).

Drugs targeting the endocrine system may represent the highest environmental potency and risk (Gunnarsson et al., 2019[112]). By way of example, the effects of the synthetic oestrogen EE2 at concentrations of 5-6 ng/L were demonstrated in a 7-year whole-lake experiment conducted in north-western Ontario, Canada. Male fish (Fathead minnow) in this study underwent feminisation, and females showed delayed ovarian development, leading to the collapse of the fish population (Kidd et al., 2007[116]). Similarly, the progestin levonorgestrel (LNG) has shown to be a highly efficient androgen in female fish. Under laboratory conditions, LNG blocked seasonal inhibition of sperm production in male fish and gave rise to a male-biased sex ratio. The latter effect was seen at a concentration of 10 ng/L, resulting in a population consisting only of male fish (Svensson et al., 2014[117]; Svensson et al., 2013[118]; Svensson et al., 2016[119]). Furthermore, laboratory exposure of LNG or EE2 to frogs have demonstrated that females lacked, or had underdeveloped, oviducts, which caused sterility (Kvarnryd et al., 2011[120]; Gyllenhammar et al., 2009[121]).

The release of hormones may also have unintended adverse effects to humans via exposure through food and drinking water. For instance, oestrogens are believed to give rise to increased risk to breast cancer in females (Moore et al., 2016[122]) and prostate cancer in men (Nelles, Hu and Prins, 2011[123]). Pregnant women, foetuses and infants are particularly at risk if continuously exposed to endocrine disruptors even at low concentrations. Potential diseases linked to endocrine disrupting chemicals include those related to the following systems: reproductive and endocrine (e.g. breast or prostate cancer, infertility, diabetes, early puberty), immune and autoimmune, cardiopulmonary (e.g. asthma or heart disease), and nervous systems (e.g. Alzheimer's disease, Parkinson's disease and attention deficit hyperactivity disorder) (WHO, 2012[115]).

Sources: (WHO, 2012[115]) (Kidd et al., 2007[116]) (Svensson et al., 2014[117]; Svensson et al., 2013[118]; Svensson et al., 2016[119]) (Kvarnryd et al., 2011[120]; Gyllenhammar et al., 2009[121]) (Moore et al., 2016[122]) (Nelles, Hu and Prins, 2011[123]) (Godfray et al., 2019[124]) (Gunnarsson et al., 2019[112]).

Table 1.4 Examples of measured effects of certain pharmaceutical residues on aquatic organisms in laboratory studies

Therapeutic group	Examples of Pharmaceutical	Impact and effected organisms	Examples of studies
Analgesics	Diclofenac, Ibuprofen	Organ damage, reduced hatching success (fish) Genotoxicity, neurotoxicity and oxidative stress (mollusk) Disruption with hormones (frog)	(Näslund et al., 2017[125]) (Mathias et al., 2018[126]) (Xia, Zheng and Zhou, 2017[127]) (Mezzelani et al., 2016[128]) (Efosa et al., 2017[129])
Antibiotics	-	Reduced growth (environmental bacteria, algae and aquatic plants)	(Roose-Amsaleg and Laverman, 2016[130]) (Guo, Boxall and Selby, 2015[131]) (Brain et al., 2008[132])
Anti-cancer	Cyclophosphamide[1], mitomycin C, fluorouracil, cisplatin, doxorubicin	Ecotoxcity, genotoxicity	(Česen et al., 2016[133]) (Zounková et al., 2007[134]) (Araújo et al., 2019[135]) (EC, 2016[136])
Antidiabetics	Metformin	Potential endocrine-disrupting effects (fish)	(Niemuth et al., 2015[137]; Crago et al., 2016[138])[2]
Anti-convulsants	Carbamazepine, phenytoin, valproic acid	Reproduction toxicity (invertebrates), development delay (fish)	(Ferrari et al., 2003[139]) (Martinez et al., 2018[140])
Antifungals	Ketoconazole, clotrimazole triclosan	Reduced growth (algae, fish), reduced algae community growth, disruption of hormones (rats)	(Vestel et al., 2016[141]) (Porsbring et al., 2009[142]) (Stoker, Gibson and Zorrilla, 2010[143])
Antihistamines	Hydroxyzine, fexofenadine, diphenhydramine	Behaviour changes, growth and feeding rate (fish) Behaviour changes and reproduction toxicity (invertebrates)	(Berninger et al., 2011[144]) (Kristofco et al., 2016[145]) (Jonsson et al., 2014[146]; Meinertz et al., 2010[147])
Antiparasitics	Ivermectin	Growth and reduced reproduction (invertebrates)	(Garric et al., 2007[148])
Beta blockers	Propranolol	Reproduction behaviour (fish), reproduction toxicity (invertebrates)	(Giltrow et al., 2009[149]) (de Oliveira et al., 2016[150])
Endocrine active pharmaceuticals	E2, EE2, levonorgestrel	Disruption with hormones causing reproduction toxicity (fish, frogs)	(Kidd et al., 2007[116]) (Kvarnryd et al., 2011[120]) (Gyllenhammar et al., 2009[121]) (Armstrong et al., 2016[151]) (Moore et al., 2016[122]) (Nelles, Hu and Prins, 2011[123])
Psychiatric dugs	Fluoxetine, sertraline, oxazepam, citalopram, chlorpromazine	Behaviour changes - feeding, boldness, activity, sociality (fish) Disruption with hormones (fish) Behaviour changes - swimming and cryptic (invertebrates) Reproduction toxicity and disruption with hormones (invertebrates)	(Brodin et al., 2013[152]; Brodin et al., 2014[113]) (Kellner et al., 2016[114]) (Schultz et al., 2011[153]) (De Castro-Català et al., 2017[154]) (Di Poi et al., 2014[155]) (Campos et al., 2016[156]) (Lazzara et al., 2012[157])

1. Transformation of Cyclophosphamide and Ifosfamide.
2. NB, Caldwell et al. (2019[158]) conclude the opposite; that metformin and its transformation product guanylurea indicate no significant environmental risk.

Economic costs are not always easily defined when it comes to the loss of biodiversity and ecosystem services caused by pharmaceuticals in the environment. However, two examples of adverse effects from diclofenac and ivermectin illustrate the costs can be substantial (see Box 1.5 and Box 1.6 respectively).

Box 1.5. Diclofenac and the collapse of vulture populations, India

Diclofenac is a non-steroidal anti-inflammatory pharmaceutical used in human and veterinary medicine. It was found to be responsible for the >90% decline in the population of three vulture species across the Indian subcontinent in the 1990s and 2000s. The birds suffered from renal failure after feeding on dead cattle previously treated with diclofenac (Green et al., 2004[159]).

As vultures are a keystone species, their population collapse has had a range of ecological, socio-economic, cultural and human impacts. Vultures have historically played an important role in environmental health, by disposing of animal and human remains. As a consequence of the loss of vultures, the dog population in India increased, which was assumed to cause increased loss of human life and health costs; dogs are the main source of rabies. The estimated medical expenses from rabid dog bites was estimated at USD 34 billion over a 14 year period (Markandya et al., 2008[160]).

In 2006, India, Pakistan and Nepal banned the veterinary usage of diclofenac, and introduced a drug called meloxicam as a substitute, to prevent further decline in vulture populations. Both the European Union and the US Environmental Protection Agency have identified diclofenac as an environmental threat. Of note, diclofenac, in this case, was not transited by water, and as such, end-of-pipe measures (e.g. wastewater treatment plant upgrades) would not have an impact on the survival of vulture populations.

Since the ban, of the three species, the White-rumped Vulture population has increased, but the Indian Vulture population continues to decline and the population size of the Slender-billed Vultures is too small to estimate a reliable trend. The main causes are thought to be the re-purposing of diclofenac for human use for livestock use, and the veterinary use of five other non-steroidal anti-inflammatory drugs (aceclofenac, carprofen, flunixin, ketoprofen and nimesulide) that are known to also be toxic to vultures (Prakash et al., 2019[161]).

Sources: (Green et al., 2004[159]) (Markandya et al., 2008[160]) (Prakash et al., 2019[161]).

Box 1.6. Potential loss of ecosystem services caused by ivermectin

Ivermectin is a parasiticide used in veterinary medicine. Risk assessments suggest risks to aquatic and terrestrial biota. The dung beetle is one of the most sensitive organisms affected by ivermectin (Liebig et al., 2010[162]). The critical ecological roles of the dung beetle include decomposition of animal waste, recycling nitrogen and control of pest habitats, with significant economic value for the cattle industry (Losey and Vaughan, 2006[163]). In the U.S., the cost of loss in ecosystem services of the dung beetle due to ivermectin in the environment was estimated at USD 380 million per year (Losey and Vaughan, 2006[163]).

Sources: (Liebig et al., 2010[162]) (Losey and Vaughan, 2006[163]).

Antibiotics are a specific group of APIs of potential ecological and human health concern for three reasons: i) all major nutrient cycles in the environment depend on the activity of bacterial communities; ii) antibiotics entering WWTPs may affect bacterial communities used for biological degradation and consequently decrease their efficiency to remove other pollutants from the water (e.g. (Schmidt, Winter and Gallert, 2012[164])) and iii) the mis- and over-use of antibiotics is an important contributing factor of antimicrobial resistance (AMR); up to 50% of the antibiotics prescribed[4] for human use are considered unnecessary (Holmes et al., 2017[165]); the number is even greater in the agriculture and aquaculture sectors where in some countries they are administered as a growth promoter and as substitute for good hygiene (Van Boeckel, 2017[18]).

Antimicrobial resistance (AMR) is a global health crisis with the potential for enormous health and economic consequences. An AMR review commissioned by the UK Prime Minister estimates that drug resistant infections cause up to 700,000 deaths each year globally, and will, if no action is undertaken, increase to 10 million per year by 2050 (Review on Antimicrobial Resistance, 2015[48]). In OECD countries, dealing with AMR complications is projected to cost up to USD 3.5 billion every year between 2015 and 2050, which corresponds to 10% of health care costs caused by communicable diseases (OECD, 2018[166]). Box 1.7 provides a brief introduction to AMR, including sources, pathways and potential the adverse effects. For more information on AMR, see OECD (2018[166]), *Stemming the Superbug Tide*.

Box 1.7. An introduction to Antimicrobial Resistance

Antimicrobial resistance (AMR) is the ability of a microbe to resist the effects of medication that could once successfully destroy or inhibit the microbe. In other words, AMR occurs when microorganisms become resistant to antibiotics, antifungals or antivirals. This phenomenon poses severe risks to global health, livelihoods and food security. Drug resistant infections are estimated to cause 700,000 deaths each year globally (IACG, 2018[167]). To put this in perspective, this is higher than the number of people dying from cancer worldwide, and may also be an underestimation because secondary effects such as avoiding surgery, were not taken into account. In Europe, North America and Australia alone, superbug infections could cost the lives of around 2.4 million over the next 30 years unless more is done to combat AMR (OECD, 2018[166]).

A continued rise in AMR is projected to lead to a reduction of 2-3.5% in Gross Domestic Product globally, with a cumulative cost of up to USD 100 trillion (Review on Antimicrobial Resistance, 2015[48]). In OECD countries, dealing with AMR complications is projected to cost up to USD 3.5 billion every year between 2015 and 2050, which corresponds to 10% of health care costs caused by communicable diseases (OECD, 2018[166]).

AMR is acquired through mutation of existing DNA or by direct exchange from other bacteria. Resistance is not a new phenomenon, bacteria have always evolved so that they can resist drugs. However, due to the misuse and overuse of antibiotics in human and veterinary medicine, AMR has increasingly become a problem as new resistance mechanisms and multi-resistant bacteria rapidly increase worldwide. Antibiotic resistance can spread across national borders via human travellers and food trade, but also via wild birds and animals (Bengtsson-Palme, Kristiansson and Larsson, 2018[168]) (Johnning et al., 2015[169]), making this a global issue.

Resistance is already high, and projected to grow even more rapidly, in low and middle-income countries. In Brazil, Indonesia and Russia, for example, between 40% and 60% of infections are already resistant, compared to an average of 17% in OECD countries. In these countries, growth of AMR rates is forecast to be 4 to 7 times higher than in OECD countries between now and 2050 (OECD, 2018[166]).

The main sources of antibiotics and AMR bacteria, evolutionary processes in the environment and exposure routes to humans are illustrated in Figure 1.5. The use of antibiotics in human health, agriculture and aquaculture practices causes releases to the environment, which itself is a vector of AMR (IACG, 2018[167]). The role of the environment in the evolution of AMR can be categorised into two processes: i) transmission of resistant genes and/or pathogens, often via faecal contamination of water (Karkman, Pärnänen and Larsson, 2019[170]); and ii) exposure of environmental bacteria to antibiotics which may develop novel (i.e. new) resistance genes which can then be transferred to pathogens

(MistraPharma, 2016[171]). Taken together, the environment becomes a reservoir for resistant genes as well as an arena for the development and spread of resistance to pathogens. Many of the resistant genes that are encountered in pathogens today originate from bacteria normally thriving in the environment. However, there is still a lack of knowledge regarding how, and under what circumstances, the environment contributes to the development of AMR (Larsson et al., 2018[172]).

Freshwater ecosystems provide an ideal setting for the acquisition and spread of AMR due to the continuous pollution by antimicrobial compounds derived from anthropogenic activities. Humans can be considered as a source of both antibiotics and antibiotic resistance genes, which may be released into the environment via WWTPs. Antibiotics are also used in veterinary practices, animal husbandry, agriculture and aquaculture.

Figure 1.5. The roles of the environment in the development of antimicrobial resistance

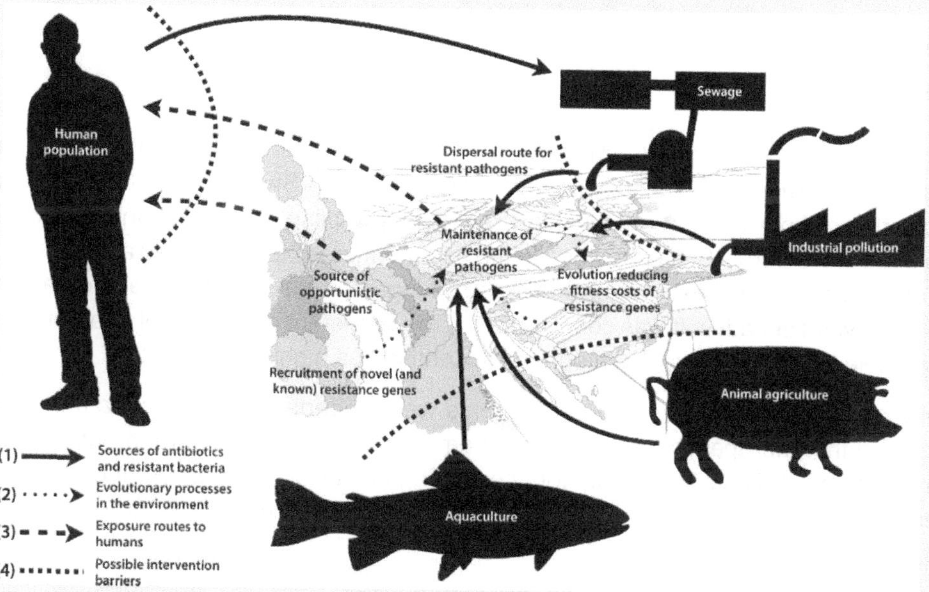

Note: Silhouettes represent common sources of antibiotics and resistance genes to the environment. The human silhouette also represents the human microbiome as a recipient of resistance genes and resistant bacteria from the environment. Dotted and dashed arrows correspond to evolutionary processes in the environment and exposure routes to humans, respectively. The dashed lines show possible points of intervention.

Source of figure: (Larsson et al., 2018[172]).

1.4.2. Pharmaceutical characteristics that affect environmental risks

It is important to not only measure the environmental concentrations of pollutants, but also whether they are toxicologically significant or insignificant, that is, whether they have the potential to cause adverse effects to human or ecosystem health at concentrations observed in the environment. When assessing the risks of pollutants in the environment, there are a number of properties of particular interest, including persistence, bioaccumulation, toxicity and mobility (Schwarzenbach et al., 2006[173]). These properties are briefly explained in the following sections.

Persistence

Persistence refers to chemical substances that withstand natural decomposition, such as biodegradation, hydrolysis or photolysis. Persistent substances increase the potential for long-term and varied effects, and a longer exposure time increases the potential for multiple contamination of the ecosystem (Kümmerer, 2009[9]). Persistent substances remain in their original form in aquatic systems for long periods of time, sometimes affecting water bodies hundreds or thousands of kilometres away from the contaminant source (Schwarzenbach et al., 2006[173]), and therefore potentially turning a pollution source into an international problem (Metz and Ingold, 2014[174]).

Pharmaceuticals that are designed to be slowly degradable, or even non-degradable, present a special risk when they enter, disseminate and persist in the environment. Such substances are referred to as 'environmentally persistent pharmaceutical pollutants (EPPPs)' (SAICM, 2015[175]). An example of an EPPP is oxazepam (see Box 1.8). Some transformation products are also hazardous (Godoy and Kummrow, 2017[33]; Česen et al., 2016[47]). In addition, there are 'pseudo-persistent' pharmaceutical pollutants, which are degradable, but continuous emissions into the environment can lead to their constant environmental presence (Daughton, 2002[176]). For example, paracetamol and ibuprofen are low persistent pharmaceuticals, but effectively behave as persistent compounds because of their continuous emission to the environment (e.g. via WWTPs), at rates faster than environmental removal (degradation) (Löffler et al., 2005[177]).

Box 1.8. Long-term persistence of oxazepam in a large freshwater lake, Sweden

Oxazepam was introduced to the pharmaceutical market in the late 1960s and is used to treat anxiety. The persistence of oxazepam in sediment was investigated in a Swedish lake (Lake Ekoln) receiving wastewater effluent from the city of Uppsala. Sediment sampling in the lake in 2013 confirmed that oxazepam used in the early 1970s remained in the sediment, despite in situ degradation processes and sediment diagenesis. The presence of oxazepam suggests that this drug can remain bioactive in sediments for several decades. It was concluded that the persistent properties of oxazepam in sediment were comparable to other chemical contaminants identified as highly persistent.

In the same study, laboratory and field experiments were undertaken on oxazepam in surface water. Results showed that therapeutic forms of oxazepam can persist over several months in cold (<5 °C) lake water free from UV light. The key factors controlling the initial degradation of oxazepam in natural waters were solar UV light, temperature and the total organic carbon of the sediment.

Source: (Klaminder et al., 2015[178]).

Mobility

Mobility refers to the potential of chemical substances to transport in soil and aquatic systems. Pharmaceuticals with certain characteristics (e.g. high water solubility, polarity and low sorption potential) are expected to be more mobile in soil and may quickly be transported to surface water or groundwater, and vice-versa. A range of factors determining the transportation of pharmaceuticals in soil and water include climate (e.g. rainfall), soil pH and the persistent properties of the pharmaceutical in question (Boxall, 2012[179]).

Bioaccumulation

Bioaccumulation refers to a compound that is incorporated into living tissue without being properly excreted or degraded within that living system. Hence, bioaccumulating substances remain in the organism, and in some cases, can also biomagnify through the food chain, meaning that the concentration increases with increasing trophic level.

Knowledge regarding pharmaceuticals in biota and bioaccumulation throughout the food web is scarce (Miller et al., 2018[180]). Reported bioaccumulation is in general higher for invertebrates compared to fish (Miller et al., 2018[180]). Differences in bioaccumulation across species may be attributed to several factors, such as lipid content of organisms, body size, life stage and respiration strategy (Miller et al., 2018[180]), but also pharmacokinetic and pharmacodynamics parameters[5] (e.g. transport system, protein binding, metabolism, partitioning into organs and tissues) (Fick et al., 2010[181]). Box 1.9 provides an example of bioaccumulation of three pharmaceuticals at multiple trophic levels in an aquatic food web.

Box 1.9. Bioaccumulation of pharmaceuticals through the aquatic food web

The uptake of five pharmaceuticals (diphenhydramine, oxazepam, trimethoprim, diclofenac and hydroxyzine) was investigated in a field study using an aquatic food-web consisting of fish (European perch) and four aquatic invertebrates (damselfly, mayfly, water louse and Ramshorn snail). The results showed that bioaccumulation of the different pharmaceuticals were species- and substance-specific.

There was no evidence that Diclofenac or Trimethoprim accumulated in the organisms (i.e. they were not detected). In contrast, diphenhydramine, oxazepam and hydroxyzine were detected in all species, with hydroxyzine generally being detected in the highest concentrations.

Mayfly larvae showed a four-fold increase of diphenhydramine, oxazepam and hydroxyzine approximately 30 days after the pharmaceuticals were added. The European perch showed an increased concentration during the study period, even though the water concentration decreased. These findings suggest that pharmaceuticals can remain bioavailable for aquatic organisms for long periods and possibly re-enter the food web at a later phase. This study demonstrates exposure and bioaccumulation; ecological effects were not part of scope. In particular, field-based experiments are needed for assessing true exposure and potential effects in natural systems.

Source: (Lagesson et al., 2016[182]).

Toxicity

Toxicity refers to the property of substances that poses a significant hazard to humans or ecosystems. Toxicity to aquatic organisms is determined by standard acute (i.e. mortality) or chronic (e.g. impaired reproduction and growth) endpoints. Acute effects are observed at concentrations orders of magnitudes higher than measured environmental concentrations, and are thus less likely to occur (Boxall, 2012[17]). Chronic effects to aquatic organisms reported for certain pharmaceuticals include growth inhibition, immobilisation, genotoxicity, reproduction toxicity, neurotoxicity and oxidative stress (see section 1.4.1 and Table 1.4).

When considering toxicity risks, there are a number of factors that can complicate assessments:

- *Mixture effects*. In real life, substances are not isolated in the environment; instead they occur mixed together and in combination with other contaminants. Even if a given substance is at a concentration too low to be harmful (i.e. below the no-observed effect concentration), when mixed in water with other chemicals, research has shown that the combined effect can be harmful (Kortenkamp et al., 2007[183]; Backhaus, 2014[184]; Godoy and Kummrow, 2017[185]). There is growing evidence that mixtures of pharmaceuticals possess a joint toxicity greater (i.e. additive effects) than individual toxicities (e.g. (Kortenkamp, Backhaus and Faust, 2009[186]).

- *Additive effects*. One type of mixture effects are additive effects; the others are antagonistic (combined effect of the mixture smaller than the sum of individual effects) or synergistic (combined effect of the mixture larger than the sum of individual effects). Current empirical knowledge suggests that pharmaceuticals that share the same mode of action (molecular targets) can act additively. For instance, the effect of three beta blockers (Propranolol, Atenolol and Metoprolol) was 10-fold more toxic to algae in terms of photosynthesis inhibition (Escher et al., 2006[187]).

Furthermore, additive toxicity of mixtures of selective serotonin reuptake inhibitors (antidepressants, such as Fluoxetine, Paroxetine, Sertraline, and Citalopram) towards aquatic invertebrates and algae has also been demonstrated (Christensen et al., 2007[188]). The progestin's LNG and norethindrone, in combination with progesterone (a natural hormone), have been shown to inhibit egg development in frogs likely acting additively, at concentrations of 1-10 ng/L, which is notable since the therapeutic plasma concentration in women is more than 1000 times higher (Säfholm et al., 2012[189]; Säfholm et al., 2015[190]). Other additive mixture effects have been documented for quinolone antibiotics (Backhaus, Scholze and Grimme, 2000[191]), oestrogens (Brian et al., 2007[192]), anti-inflammatory drugs (Cleuvers, 2003[193]) and anticancer drugs (Elersek et al., 2016[194]). In addition, there is evidence that pharmaceuticals not sharing the same mode of action (i.e. belonging to different therapeutic groups with different molecular targets) can also have additive effects (Runnalls et al., 2015[195]) (Thrupp et al., 2018[196]) (Neale, Leusch and Escher, 2017[197]) (Yamagishi, Horie and Tatarazako, 2017[198]).

- *Effects of metabolites and transformation products.* In the environment, transformation and degradation reactions alter the mobility, persistence and fate of the pharmaceutical residues (Kümmerer, 2009[9]) (Weber et al., 2014[3]). Metabolites and transformation products represent a large amount of the total pharmaceuticals load. This is of concern since these compounds are rarely evaluated for their toxicity, occurrence or fate, and some evidence suggests that the ecotoxicity of some degradation products can have higher toxicity effects than that of their parent compounds (Godoy and Kummrow, 2017[185]). For example, the transformation products of ranitidine (a histamine) and of Naproxen (a nonsteroidal anti-inflammatory) have been found to be more toxic than the parent compounds, both for acute and chronic values (Isidori et al., 2009[199]; Isidori et al., 2005[200]). In the Netherlands, the measured presence of transformation products with similar pharmacological activities and concentration levels as their parent compounds illustrates the relevance of monitoring transformation products (and metabolites), and including these in environmental risk assessments (de Jongh et al., 2012[67]).

- *Multiple routes of exposure.* Ecosystems and humans may be continuously exposed via a number of pathways (see section 1.3.2) of low-dose mixtures that can have additive effects.

The main challenges faced by regulators is how to determine the degree to which humans and ecosystems are exposed to pharmaceuticals, what interactions may occur, and what specific human and ecosystem health impacts are associated with chemical mixtures (WHO, 2017[201]). Given the large number of pharmaceuticals on the market, and possible mixture combinations (with other pharmaceuticals and other chemicals), exposure pathways, additive effects and transformation products, it can easily be said that assessing the risk of all possible scenarios would not be feasible. Chapter 2 will consider various monitoring techniques and strategies to assess potential risks.

References

Adeel, M. et al. (2017), "Environmental impact of estrogens on human, animal and plant life: A critical review", *Environment International*, Vol. 99, pp. 107-119, http://dx.doi.org/10.1016/j.envint.2016.12.010. [46]

Alexandratos, N. and J. Bruinsma (2012), *World Agriculture towards 2030/2050: the 2012 revision*, Food and Agriculture Organization of the United Nations, http://www.fao.org/economic/esa. [87]

Araújo, A. et al. (2019), "Anti-cancer drugs in aquatic environment can cause cancer: Insight about mutagenicity in tadpoles", *Science of The Total Environment*, Vol. 650, pp. 2284-2293, http://dx.doi.org/10.1016/J.SCITOTENV.2018.09.373. [135]

Armstrong, B. et al. (2016), "Reproductive effects in fathead minnows (Pimphales promelas) following a 21 d exposure to 17α-ethinylestradiol", *Chemosphere*, Vol. 144, pp. 366-373, http://dx.doi.org/10.1016/j.chemosphere.2015.08.078. [151]

Arnold, K. et al. (2014), "Medicating the environment: assessing risks of pharmaceuticals to wildlife and ecosystems", *Philosophical Transactions of the Royal Society B: Biological Sciences*, Vol. 369/1656, http://dx.doi.org/10.1098/RSTB.2013.0569. [8]

Arpin-Pont, L. et al. (2016), "Occurrence of PPCPs in the marine environment: a review", *Environmental Science and Pollution Research*, Vol. 23/6, pp. 4978-4991, http://dx.doi.org/10.1007/s11356-014-3617-x. [72]

aus der Beek, T. et al. (2016), "Pharmaceuticals in the environment-Global occurrences and perspectives", *Environmental Toxicology and Chemistry*, Vol. 35/4, pp. 823-835, http://dx.doi.org/10.1002/etc.3339. [5]

Azuma, T. et al. (2016), "Detection of pharmaceuticals and phytochemicals together with their metabolites in hospital effluents in Japan, and their contribution to sewage treatment plant influents", *Science of the Total Environment*, Vol. 548-549, pp. 189-197, http://dx.doi.org/10.1016/j.scitotenv.2015.12.157. [28]

Backhaus, T. (2014), "Medicines, shaken and stirred: A critical review on the ecotoxicology of pharmaceutical mixtures", *Philosophical Transactions of the Royal Society B: Biological Sciences*, Vol. 369/1656, http://dx.doi.org/10.1098/rstb.2013.0585. [184]

Backhaus, T., M. Scholze and L. Grimme (2000), "The single substance and mixture toxicity of quinolones to the bioluminescent bacterium Vibrio fischeri", *Aquatic Toxicology*, Vol. 49/1-2, pp. 49-61, http://dx.doi.org/10.1016/S0166-445X(99)00069-7. [191]

Baken, K. et al. (2018), "Toxicological risk assessment and prioritization of drinking water relevant contaminants of emerging concern", *Environment International*, Vol. 118, pp. 293-303, http://dx.doi.org/10.1016/J.ENVINT.2018.05.006. [102]

Barnes, K. et al. (2004), "Pharmaceuticals and other organic waste water contaminants within a leachate plume downgradient of a municipal landfill", *Ground Water Monitoring and Remediation*, Vol. 24/2, pp. 119-126, http://dx.doi.org/10.1111/j.1745-6592.2004.tb00720.x. [41]

Barnes, K. et al. (2008), "A national reconnaissance of pharmaceuticals and other organic wastewater contaminants in the United States - I) Groundwater", *Science of the Total Environment*, Vol. 402/2-3, pp. 192-200, http://dx.doi.org/10.1016/j.scitotenv.2008.04.028. [71]

Behera, S. et al. (2011), "Occurrence and removal of antibiotics, hormones and several other pharmaceuticals in wastewater treatment plants of the largest industrial city of Korea", *Science of The Total Environment*, Vol. 409/20, pp. 4351-4360, http://dx.doi.org/10.1016/J.SCITOTENV.2011.07.015. [35]

Belloni, A., D. Morgan and V. Paris (2016), "Pharmaceutical Expenditure And Policies: Past Trends And Future Challenges", *OECD Health Working Papers*, No. 87, OECD Publishing, Paris, http://dx.doi.org/10.1787/5jm0q1f4cdq7-en. [84]

Bengtsson-Palme, J., E. Kristiansson and D. Larsson (2018), "Environmental factors influencing the development and spread of antibiotic resistance", *FEMS Microbiology Reviews*, Vol. 42/1, pp. 68-80, http://dx.doi.org/10.1093/femsre/fux053. [168]

Benotti, M. et al. (2009), "Pharmaceuticals and endocrine disrupting compounds in U.S. drinking water", *Environmental Science and Technology*, Vol. 43/3, pp. 597-603, http://dx.doi.org/10.1021/es801845a. [65]

Bercu, J. et al. (2008), "Human health risk assessments for three neuropharmaceutical compounds in surface waters", *Regulatory Toxicology and Pharmacology*, Vol. 50/3, pp. 420-427, http://dx.doi.org/10.1016/j.yrtph.2008.01.014. [98]

Bernhardt, E., E. Rosi and M. Gessner (2017), "Synthetic chemicals as agents of global change", *Frontiers in Ecology and the Environment*, Vol. 15/2, pp. 84-90, http://dx.doi.org/10.1002/fee.1450. [7]

Berninger, J. et al. (2011), "Effects of the antihistamine diphenhydramine on selected aquatic organisms", *Environmental Toxicology and Chemistry*, Vol. 30/9, pp. 2065-2072, http://dx.doi.org/10.1002/etc.590. [144]

Bernot, M., L. Smith and J. Frey (2013), "Human and veterinary pharmaceutical abundance and transport in a rural central Indiana stream influenced by confined animal feeding operations (CAFOs)", *Science of the Total Environment*, Vol. 445-446, pp. 219-230, http://dx.doi.org/10.1016/j.scitotenv.2012.12.039. [62]

BIO Intelligence Service (2013), *Study on the environmental risks of medicinal products.* [107]

BIO Intelligence Service (2013), *Study on the environmental risks of medicinal products, Final Report prepared for Executive Agency for Health and Consumers*, BIO Intelligence Service, Paris, https://ec.europa.eu/health/sites/health/files/files/environment/study_environment.pdf. [20]

Boxall, A. (2012), *New and Emerging Water Pollutants arising from Agriculture*, OECD, Paris, https://www.oecd.org/tad/sustainable-agriculture/49848768.pdf. [17]

Boxall, A. (2012), *New and Emerging Water Pollutants arising from Agriculture*, OECD publishing, Paris, https://www.oecd.org/tad/sustainable-agriculture/49848768.pdf. [179]

Boxall, A. (2010), "Veterinary Medicines and the Environment", Springer, Berlin, Heidelberg, http://dx.doi.org/10.1007/978-3-642-10324-7_12. [43]

Boxall, A. et al. (2009), "Impacts of Climate Change on Indirect Human Exposure to Pathogens and Chemicals from Agriculture", *Environmental Health Perspectives*, Vol. 117/4, pp. 508-514, http://dx.doi.org/10.1289/ehp.0800084. [95]

Brain, R. et al. (2008), "Aquatic plants exposed to pharmaceuticals: effects and risks", *Reviews of environmental contamination and toxicology*, Vol. 192, pp. 67-115, http://www.ncbi.nlm.nih.gov/pubmed/18020304 (accessed on 27 June 2019). [132]

Brian, J. et al. (2007), "Evidence of estrogenic mixture effects on the reproductive performance of fish", *Environmental Science and Technology*, Vol. 41/1, pp. 337-344, http://dx.doi.org/10.1021/es0617439. [192]

Brodin, T. et al. (2013), "Dilute concentrations of a psychiatric drug alter behavior of fish from natural populations", *Science*, Vol. 339/6121, pp. 814-815, http://dx.doi.org/10.1126/science.1226850. [152]

Brodin, T. et al. (2014), "Ecological effects of pharmaceuticals in aquatic systems—impacts through behavioural alterations", *Philosophical Transactions of the Royal Society B: Biological Sciences*, Vol. 369/1656, http://dx.doi.org/10.1098/rstb.2013.0580. [113]

Burns, E. et al. (2018), *Application of prioritization approaches to optimize environmental monitoring and testing of pharmaceuticals*, Taylor and Francis Inc., http://dx.doi.org/10.1080/10937404.2018.1465873. [1]

Caldwell, D. et al. (2019), "Environmental risk assessment of metformin and its transformation product guanylurea: II. Occurrence in surface waters of Europe and the United States and derivation of predicted no-effect concentrations", *Chemosphere*, Vol. 216, pp. 855-865, http://dx.doi.org/10.1016/J.CHEMOSPHERE.2018.10.038. [158]

Campos, B. et al. (2016), "Depressing Antidepressant: Fluoxetine Affects Serotonin Neurons Causing Adverse Reproductive Responses in Daphnia magna", *Environmental Science and Technology*, Vol. 50/11, pp. 6000-6007, http://dx.doi.org/10.1021/acs.est.6b00826. [156]

Carter, L. et al. (2014), "Fate and Uptake of Pharmaceuticals in Soil–Plant Systems", *Journal of Agricultural and Food Chemistry*, Vol. 62/4, pp. 816-825, http://dx.doi.org/10.1021/jf404282y. [45]

Cavicchioli, R. et al. (2019), "Scientists' warning to humanity: microorganisms and climate change", *Nature Reviews Microbiology*, p. 1, http://dx.doi.org/10.1038/s41579-019-0222-5. [91]

Česen, M. et al. (2016), "Ecotoxicity and genotoxicity of cyclophosphamide, ifosfamide, their metabolites/transformation products and their mixtures", *Environmental Pollution*, Vol. 210, pp. 192-201, http://dx.doi.org/10.1016/j.envpol.2015.12.017. [133]

Christensen, A. et al. (2007), "Mixture and single-substance toxicity of selective serotonin reuptake inhibitors toward algae and crustaceans", *Environmental Toxicology and Chemistry*, Vol. 26/1, pp. 85-91, http://dx.doi.org/10.1897/06-219R.1. [188]

Civity (2017), *Pharmaceutical usage in the context of demographic change: The significance of growing medication consumption in Germany for raw water resources*, Civity Management Consultants. [86]

Cleuvers, M. (2003), "Aquatic ecotoxicity of pharmaceuticals including the assessment of combination effects", *Toxicology Letters*, Vol. 142/3, pp. 185-194, http://dx.doi.org/10.1016/S0378-4274(03)00068-7. [193]

Collier, A. (2007), "Pharmaceutical Contaminants in Potable Water: Potential Concerns for Pregnant Women and Children", *EcoHealth*, Vol. 4/2, pp. 164-171, http://dx.doi.org/10.1007/s10393-007-0105-5. [104]

Comber, S. et al. (2018), "Active pharmaceutical ingredients entering the aquatic environment from wastewater treatment works: A cause for concern?", *Science of The Total Environment*, Vol. 613-614, pp. 538-547, http://dx.doi.org/10.1016/J.SCITOTENV.2017.09.101. [37]

Crago, J. et al. (2016), "Age-dependent effects in fathead minnows from the anti-diabetic drug metformin", *General and Comparative Endocrinology*, Vol. 232, pp. 185-190, http://dx.doi.org/10.1016/j.ygcen.2015.12.030. [138]

Cunningham, V., S. Binks and M. Olson (2009), "Human health risk assessment from the presence of human pharmaceuticals in the aquatic environment", *Regulatory Toxicology and Pharmacology*, Vol. 53/1, pp. 39-45, http://dx.doi.org/10.1016/j.yrtph.2008.10.006. [99]

Daughton, C. (2002), "Environmental stewardship and drugs as pollutants", *The Lancet*, Vol. 360/9339, pp. 1035-1036, http://dx.doi.org/10.1016/S0140-6736(02)11176-7. [176]

Daughton, C. and I. Ruhoy (2009), "Environmental footprint of pharmaceuticals: The significance of factors beyond direct excretion to sewers", *Environmental Toxicology and Chemistry*, Vol. 28/12, pp. 2495-2521, http://dx.doi.org/10.1897/08-382.1. [30]

Daughton, C. and T. Ternes (1999), "Pharmaceuticals and personal care products in the environment: Agents of subtle change?", *Environmental Health Perspectives*, Vol. 107/SUPPL. 6, pp. 907-938, http://dx.doi.org/10.1289/ehp.99107s6907. [66]

De Castro-Català, N. et al. (2017), "Evidence of low dose effects of the antidepressant fluoxetine and the fungicide prochloraz on the behavior of the keystone freshwater invertebrate Gammarus pulex", *Environmental Pollution*, Vol. 231, pp. 406-414, http://dx.doi.org/10.1016/j.envpol.2017.07.088. [154]

de Jongh, C. et al. (2012), "Screening and human health risk assessment of pharmaceuticals and their transformation products in Dutch surface waters and drinking water", *Science of the Total Environment*, Vol. 427-428, pp. 70-77, http://dx.doi.org/10.1016/j.scitotenv.2012.04.010. [67]

de Oliveira, L. et al. (2016), "Acute and chronic ecotoxicological effects of four pharmaceuticals drugs on cladoceran Daphnia magna", *Drug and Chemical Toxicology*, Vol. 39/1, pp. 13-21, http://dx.doi.org/10.3109/01480545.2015.1029048. [150]

Deblonde, T., C. Cossu-Leguille and P. Hartemann (2011), "Emerging pollutants in wastewater: A review of the literature", *International Journal of Hygiene and Environmental Health*, Vol. 214/6, pp. 442-448, http://dx.doi.org/10.1016/j.ijheh.2011.08.002. [36]

Di Poi, C. et al. (2014), "Cryptic and biochemical responses of young cuttlefish Sepia officinalis exposed to environmentally relevant concentrations of fluoxetine", *Aquatic Toxicology*, Vol. 151, pp. 36-45, http://dx.doi.org/10.1016/j.aquatox.2013.12.026. [155]

Donnachie, R., A. Johnson and J. Sumpter (2016), "A rational approach to selecting and ranking some pharmaceuticals of concern for the aquatic environment and their relative importance compared with other chemicals", *Environmental Toxicology and Chemistry*, http://dx.doi.org/10.1002/etc.3165. [96]

EC (2016), *Fate and effects of cytostatic pharmaceuticals in the environment and the identification of biomarkers for and improved risk assessment on environmental exposure*, European Commission, Brussels, https://cordis.europa.eu/project/rcn/96703/brief/en. [136]

Efosa, N. et al. (2017), "Diclofenac can exhibit estrogenic modes of action in male Xenopus laevis, and affects the hypothalamus-pituitary-gonad axis and mating vocalizations", *Chemosphere*, Vol. 173, pp. 69-77, http://dx.doi.org/10.1016/j.chemosphere.2017.01.030. [129]

Elersek, T. et al. (2016), "Toxicity of the mixture of selected antineoplastic drugs against aquatic primary producers", *Environmental Science and Pollution Research*, Vol. 23/15, pp. 14780-14790, http://dx.doi.org/10.1007/s11356-015-6005-2. [194]

Ericson, H., G. Thorsén and L. Kumblad (2010), "Physiological effects of diclofenac, ibuprofen and propranolol on Baltic Sea blue mussels", *Aquatic Toxicology*, Vol. 99/2, pp. 223-231, http://dx.doi.org/10.1016/J.AQUATOX.2010.04.017. [78]

Escher, B. et al. (2006), "Comparative ecotoxicological hazard assessment of beta-blockers and their human metabolites using a mode-of-action-based test battery and a QSAR approach", *Environmental Science and Technology*, Vol. 40/23, pp. 7402-7408, http://dx.doi.org/10.1021/es052572v. [187]

European Commission (2006), *REGULATION (EC) No 1907/2006 OF THE EUROPEAN PARLIAMENT AND OF THE COUNCIL of 18 December 2006 concerning the Registration, Evaluation, Authorisation and Restriction of Chemicals (REACH)*, https://eur-lex.europa.eu/legal-content/EN/TXT/PDF/?uri=CELEX:32006R1907&from=EN. [202]

Fabbri, E. and S. Franzellitti (2016), "Human pharmaceuticals in the marine environment: Focus on exposure and biological effects in animal species", *Environmental Toxicology and Chemistry*, Vol. 35/4, pp. 799-812, http://dx.doi.org/10.1002/etc.3131. [74]

FAO (2018), *More people, more food, worse water? A global review of water pollution from agriculture*, Food and Agriculture Organization of the United Nations, Rome, https://reliefweb.int/sites/reliefweb.int/files/resources/CA0146EN.pdf. [44]

FAO (2016), *The State of World Fisheries and Aquaculture 2016: Contributing to food security and nutrition for all*, Food and Agriculture Organization of the United Nations, Rome. [50]

Ferrari, B. et al. (2003), "Ecotoxicological impact of pharmaceuticals found in treated wastewaters: Study of carbamazepine, clofibric acid, and diclofenac", *Ecotoxicology and Environmental Safety*, Vol. 55/3, pp. 359-370, http://dx.doi.org/10.1016/S0147-6513(02)00082-9. [139]

Fick, J. et al. (2010), "Predicted critical environmental concentrations for 500 pharmaceuticals", *Regulatory Toxicology and Pharmacology*, Vol. 58/3, pp. 516-523, http://dx.doi.org/10.1016/j.yrtph.2010.08.025. [181]

Fick, J. et al. (2009), "Contamination of surface, ground, and drinking water from pharmaceutical production", *Environmental Toxicology and Chemistry*, Vol. 28/12, pp. 2522-2527, http://dx.doi.org/10.1897/09-073.1. [23]

Focazio, M. et al. (2008), "A national reconnaissance for pharmaceuticals and other organic wastewater contaminants in the United States - II) Untreated drinking water sources", *Science of the Total Environment*, Vol. 402/2-3, pp. 201-216, http://dx.doi.org/10.1016/j.scitotenv.2008.02.021. [69]

Forrest, F. et al. (2011), "A scoping study of livestock antimicrobials in agricultural streams of Alberta", *Canadian Water Resources Journal*, Vol. 36/1, pp. 1-16, http://dx.doi.org/10.4296/cwrj3601001. [63]

Gardner, M. et al. (2012), "The significance of hazardous chemicals in wastewater treatment works effluents", *Science of The Total Environment*, Vol. 437, pp. 363-372, http://dx.doi.org/10.1016/J.SCITOTENV.2012.07.086. [38]

Garric, J. et al. (2007), "Effects of the parasiticide ivermectin on the cladoceran Daphnia magna and the green alga Pseudokirchneriella subcapitata", *Chemosphere*, Vol. 69/6, pp. 903-910, http://dx.doi.org/10.1016/j.chemosphere.2007.05.070. [148]

Gaw, S., K. Thomas and T. Hutchinson (2014), "Sources, impacts and trends of pharmaceuticals in the marine and coastal environment", *Philosophical Transactions of the Royal Society B: Biological Sciences*, Vol. 369/1656, pp. 20130572-20130572, http://dx.doi.org/10.1098/rstb.2013.0572. [73]

Giltrow, E. et al. (2009), "Chronic effects assessment and plasma concentrations of the β-blocker propranolol in fathead minnows (Pimephales promelas)", *Aquatic Toxicology*, Vol. 95/3, pp. 195-202, http://dx.doi.org/10.1016/j.aquatox.2009.09.002. [149]

Godfray, H. et al. (2019), "A restatement of the natural science evidence base on the effects of endocrine disrupting chemicals on wildlife", *Proceedings of the Royal Society B: Biological Sciences*, Vol. 286/1897, p. 20182416, http://dx.doi.org/10.1098/rspb.2018.2416. [124]

Godoy, A. and F. Kummrow (2017), "What do we know about the ecotoxicology of pharmaceutical and personal care product mixtures? A critical review", *Critical Reviews in Environmental Science and Technology*, Vol. 47/16, pp. 1453-1496, http://dx.doi.org/10.1080/10643389.2017.1370991. [185]

Gonzalez-Rey, M. and M. Bebianno (2014), "Effects of non-steroidal anti-inflammatory drug (NSAID) diclofenac exposure in mussel Mytilus galloprovincialis", *Aquatic Toxicology*, Vol. 148, pp. 221-230, http://dx.doi.org/10.1016/J.AQUATOX.2014.01.011. [80]

Green, R. et al. (2004), "Diclofenac poisoning as a cause of vulture population declines across the Indian subcontinent", *Journal of Applied Ecology*, Vol. 41/5, pp. 793-800, http://dx.doi.org/10.1111/j.0021-8901.2004.00954.x. [110]

Green, R. et al. (2004), "Diclofenac poisoning as a cause of vulture population declines across the Indian subcontinent", *Journal of Applied Ecology*, Vol. 41/5, pp. 793-800, http://dx.doi.org/10.1111/j.0021-8901.2004.00954.x. [159]

Guler, Y. and A. Ford (2010), "Anti-depressants make amphipods see the light", *Aquatic Toxicology*, Vol. 99/3, pp. 397-404, http://dx.doi.org/10.1016/J.AQUATOX.2010.05.019. [77]

Gunnarsson, L. et al. (2019), "Pharmacology beyond the patient – The environmental risks of human drugs", *Environment International*, Vol. 129, pp. 320-332, http://dx.doi.org/10.1016/J.ENVINT.2019.04.075. [112]

Guo, J., A. Boxall and K. Selby (2015), "Do pharmaceuticals pose a threat to primary producers?", *Critical Reviews in Environmental Science and Technology*, Vol. 45/23, http://dx.doi.org/10.1080/10643389.2015.1061873. [131]

Gyllenhammar, I. et al. (2009), "Reproductive toxicity in Xenopus tropicalis after developmental exposure to environmental concentrations of ethynylestradiol", *Aquatic Toxicology*, Vol. 91/2, pp. 171-178, http://dx.doi.org/10.1016/j.aquatox.2008.06.019. [121]

Heberer, T. and D. Feldmann (2005), "Contribution of effluents from hospitals and private households to the total loads of diclofenac and carbamazepine in municipal sewage effluents - Modeling versus measurements", *Journal of Hazardous Materials*, Vol. 122/3, pp. 211-218, http://dx.doi.org/10.1016/j.jhazmat.2005.03.007. [14]

Hollender, J. et al. (2009), "Elimination of Organic Micropollutants in a Municipal Wastewater Treatment Plant Upgraded with a Full-Scale Post-Ozonation Followed by Sand Filtration", *Environmental Science & Technology*, Vol. 43/20, pp. 7862-7869, http://dx.doi.org/10.1021/es9014629. [34]

Holmes, D. et al. (2017), "A geospatial approach for identifying and exploring potential naturalwater storage sites", *Water (Switzerland)*, Vol. 9/8, http://dx.doi.org/10.3390/w9080585. [165]

Houtman, C. et al. (2014), "Human health risk assessment of the mixture of pharmaceuticals in Dutch drinking water and its sources based on frequent monitoring data", *Science of the Total Environment*, Vol. 496, pp. 54-62, http://dx.doi.org/10.1016/j.scitotenv.2014.07.022. [103]

IACG (2018), "Future Global Governance for Antimicrobial Resistance IACG Discussion Paper July 2018", Interagency Coordination Group on Antimicrobial Resistance, IACG, http://www.who.int/antimicrobial-resistance/interagency-coordination-group/IACG_Future_global_governance_for_AMR_120718.pdf?ua=1. [167]

Isidori, M. et al. (2005), "Ecotoxicity of naproxen and its phototransformation products", *Science of The Total Environment*, Vol. 348/1-3, pp. 93-101, http://dx.doi.org/10.1016/J.SCITOTENV.2004.12.068. [200]

Isidori, M. et al. (2009), "Effects of ranitidine and its photoderivatives in the aquatic environment", *Environment International*, Vol. 35/5, pp. 821-825, http://dx.doi.org/10.1016/J.ENVINT.2008.12.002. [199]

Jaimes-Correa, J., D. Snow and S. Bartelt-Hunt (2015), "Seasonal occurrence of antibiotics and a beta agonist in an agriculturally-intensive watershed", *Environmental Pollution*, Vol. 205, pp. 87-96, http://dx.doi.org/10.1016/j.envpol.2015.05.023. [61]

Johnning, A. et al. (2015), "Quinolone resistance mutations in the faecal microbiota of Swedish travellers to India", *BMC Microbiology*, Vol. 15/1, http://dx.doi.org/10.1186/s12866-015-0574-6. [169]

Johnning, A. et al. (2013), "Acquired genetic mechanisms of a multiresistant bacterium isolated from a treatment plant receiving wastewater from antibiotic production", *Applied and Environmental Microbiology*, Vol. 79/23, pp. 7256-7263, http://dx.doi.org/10.1128/AEM.02141-13. [25]

Johnson, A. et al. (2008), "Do cytotoxic chemotherapy drugs discharged into rivers pose a risk to the environment and human health? An overview and UK case study", *Journal of Hydrology*, Vol. 348/1-2, pp. 167-175, http://dx.doi.org/10.1016/J.JHYDROL.2007.09.054. [106]

Johnson, A. et al. (2008), "Do cytotoxic chemotherapy drugs discharged into rivers pose a risk to the environment and human health? An overview and UK case study", *Journal of Hydrology*, Vol. 348/1-2, pp. 167-175, http://dx.doi.org/10.1016/j.jhydrol.2007.09.054. [100]

Jones, A. et al. (2019), "Bluetongue risk under future climates", *Nature Climate Change*, Vol. 9/2, pp. 153-157, http://dx.doi.org/10.1038/s41558-018-0376-6. [94]

Jonsson, M. et al. (2014), "Antihistamines and aquatic insects: Bioconcentration and impacts on behavior in damselfly larvae (Zygoptera)", *Science of the Total Environment*, Vol. 472, pp. 108-111, http://dx.doi.org/10.1016/j.scitotenv.2013.10.104. [146]

Karkman, A., K. Pärnänen and D. Larsson (2019), "Fecal pollution can explain antibiotic resistance gene abundances in anthropogenically impacted environments", *Nature Communications*, Vol. 10/1, p. 80, http://dx.doi.org/10.1038/s41467-018-07992-3. [170]

Kasprzyk-Hordern, B., R. Dinsdale and A. Guwy (2008), "The occurrence of pharmaceuticals, personal care products, endocrine disruptors and illicit drugs in surface water in South Wales, UK", *Water Research*, Vol. 42/13, pp. 3498-3518, http://dx.doi.org/10.1016/j.watres.2008.04.026. [59]

Kellner, M. et al. (2016), "Waterborne citalopram has anxiolytic effects and increases locomotor activity in the three-spine stickleback (Gasterosteus aculeatus)", *Aquatic Toxicology*, Vol. 173, pp. 19-28, http://dx.doi.org/10.1016/j.aquatox.2015.12.026. [114]

Khetan, S. and T. Collins (2007), "Human pharmaceuticals in the aquatic environment: A challenge to green chemisty", *Chemical Reviews*, Vol. 107/6, pp. 2319-2364, http://dx.doi.org/10.1021/cr020441w. [6]

Kidd, K. et al. (2007), "Collapse of a fish population after exposure to a synthetic estrogen", *Proceedings of the National Academy of Sciences of the United States of America*, Vol. 104/21, pp. 8897-8901, http://dx.doi.org/10.1073/pnas.0609568104. [116]

Klaminder, J. et al. (2015), "Long-Term Persistence of an Anxiolytic Drug (Oxazepam) in a Large Freshwater Lake", *Environmental Science and Technology*, Vol. 49/17, pp. 10406-10412, http://dx.doi.org/10.1021/acs.est.5b01968. [178]

Kortenkamp, A., T. Backhaus and M. Faust (2009), *State of the Art Report on Mixture Toxicity*, The School of Pharmacy, University of London, http://ec.europa.eu/environment/chemicals/effects/pdf/report_mixture_toxicity.pdf. [186]

Kortenkamp, A. et al. (2007), "Low-level exposure to multiple chemicals: Reason for human health concerns?", *Environmental Health Perspectives*, Vol. 115/SUPPL1, pp. 106-114, http://dx.doi.org/10.1289/ehp.9358. [183]

Kostich, M., A. Batt and J. Lazorchak (2014), "Concentrations of prioritized pharmaceuticals in effluents from 50 large wastewater treatment plants in the US and implications for risk estimation", *Environmental Pollution*, Vol. 184, pp. 354-359, http://dx.doi.org/10.1016/J.ENVPOL.2013.09.013. [101]

Kot-Wasik, A., A. Jakimska and M. Śliwka-Kaszyńska (2016), "Occurrence and seasonal variations of 25 pharmaceutical residues in wastewater and drinking water treatment plants", *Environmental Monitoring and Assessment*, Vol. 188/12, http://dx.doi.org/10.1007/s10661-016-5637-0. [57]

Kristiansson, E. et al. (2011), "Pyrosequencing of antibiotic-contaminated river sediments reveals high levels of resistance and gene transfer elements", *PLoS ONE*, Vol. 6/2, http://dx.doi.org/10.1371/journal.pone.0017038. [22]

Kristofco, L. et al. (2016), "Age matters: Developmental stage of Danio rerio larvae influences photomotor response thresholds to diazinion or diphenhydramine", *Aquatic Toxicology*, Vol. 170, pp. 344-354, http://dx.doi.org/10.1016/J.AQUATOX.2015.09.011. [145]

Kümmerer, K. (2009), "The presence of pharmaceuticals in the environment due to human use – present knowledge and future challenges", *Journal of Environmental Management*, Vol. 90/8, pp. 2354-2366, http://dx.doi.org/10.1016/J.JENVMAN.2009.01.023. [9]

Küster, A. and N. Adler (2014), "Pharmaceuticals in the environment: Scientific evidence of risks and its regulation", *Philosophical Transactions of the Royal Society B: Biological Sciences*, Vol. 369/1656, http://dx.doi.org/10.1098/rstb.2013.0587. [111]

Kvarnryd, M. et al. (2011), "Early life progestin exposure causes arrested oocyte development, oviductal agenesis and sterility in adult Xenopus tropicalis frogs", *Aquatic Toxicology*, Vol. 103/1-2, pp. 18-24, http://dx.doi.org/10.1016/j.aquatox.2011.02.003. [120]

Lacorte, S. et al. (2018), "Pharmaceuticals released from senior residences: occurrence and risk evaluation", *Environmental Science and Pollution Research*, Vol. 25/7, pp. 6095-6106, http://dx.doi.org/10.1007/s11356-017-9755-1. [29]

Lagesson, A. et al. (2016), "Bioaccumulation of five pharmaceuticals at multiple trophic levels in an aquatic food web - Insights from a field experiment", *Science of the Total Environment*, Vol. 568, pp. 208-215, http://dx.doi.org/10.1016/j.scitotenv.2016.05.206. [182]

Lapworth, D. et al. (2012), "Emerging organic contaminants in groundwater: A review of sources, fate and occurrence", *Environmental Pollution*, Cited By (since 1996):98
Export Date: 20 November 2014, pp. 287-303, http://www.scopus.com/inward/record.url?eid=2-s2.0-84856446159&partnerID=40&md5=3907089c51794cbf32d82f845fbea5b2. [11]

Larsson, D. (2014), "Pollution from drug manufacturing: review and perspectives", *Philosophical transactions of the Royal Society of London. Series B, Biological sciences*, Vol. 369/1656, http://dx.doi.org/10.1098/rstb.2013.0571. [12]

Larsson, D. (2014), "Pollution from drug manufacturing: Review and perspectives", *Philosophical Transactions of the Royal Society B: Biological Sciences*, Vol. 369/1656, http://dx.doi.org/10.1098/rstb.2013.0571. [21]

Larsson, D. et al. (2018), "Critical knowledge gaps and research needs related to the environmental dimensions of antibiotic resistance", *Environment International*, Vol. 117, pp. 132-138, http://dx.doi.org/10.1016/j.envint.2018.04.041. [172]

Larsson, D., C. de Pedro and N. Paxeus (2007), "Effluent from drug manufactures contains extremely high levels of pharmaceuticals", *Journal of Hazardous Materials*, Vol. 148/3, pp. 751-755, http://dx.doi.org/10.1016/j.jhazmat.2007.07.008. [24]

Larsson, D., C. de Pedro and N. Paxeus (2007), "Effluent from drug manufactures contains extremely high levels of pharmaceuticals", *Journal of Hazardous Materials*, Vol. 148/3, pp. 751-755, http://dx.doi.org/10.1016/j.jhazmat.2007.07.008. [19]

Lazzara, R. et al. (2012), "Low environmental levels of fluoxetine induce spawning and changes in endogenous estradiol levels in the zebra mussel Dreissena polymorpha", *Aquatic Toxicology*, Vol. 106-107, pp. 123-130, http://dx.doi.org/10.1016/j.aquatox.2011.11.003. [157]

Liebig, M. et al. (2010), "Environmental risk assessment of ivermectin: A case study", *Integrated Environmental Assessment and Management*, Vol. 6/SUPPL. 1, pp. 567-587, http://dx.doi.org/10.1002/ieam.96. [162]

Lindholm-Lehto, P. et al. (2016), "Widespread occurrence and seasonal variation of pharmaceuticals in surface waters and municipal wastewater treatment plants in central Finland", *Environmental Science and Pollution Research*, Vol. 23/8, pp. 7985-7997, http://dx.doi.org/10.1007/s11356-015-5997-y. [53]

Lissemore, L. et al. (2006), "An exposure assessment for selected pharmaceuticals within a watershed in Southern Ontario", *Chemosphere*, Vol. 64/5, pp. 717-729, http://dx.doi.org/10.1016/j.chemosphere.2005.11.015. [64]

Löffler, D. et al. (2005), "Environmental Fate of Pharmaceuticals in Water/Sediment Systems", *Environmental Science & Technology*, Vol. 39/14, pp. 5209-5218, http://dx.doi.org/10.1021/es0484146. [177]

Losey, J. and M. Vaughan (2006), "The economic value of ecological services provided by insects", *BioScience*, Vol. 56/4, pp. 311-323, http://dx.doi.org/10.1641/0006-3568(2006)56[311:TEVOES]2.0.CO;2. [163]

MacFadden, D. et al. (2018), "Antibiotic resistance increases with local temperature", *Nature Climate Change*, Vol. 8/6, pp. 510-514, http://dx.doi.org/10.1038/s41558-018-0161-6. [93]

Madikizela, L., N. Tavengwa and L. Chimuka (2017), "Status of pharmaceuticals in African water bodies: Occurrence, removal and analytical methods", *Journal of Environmental Management*, Vol. 193, pp. 211-220, http://dx.doi.org/10.1016/j.jenvman.2017.02.022. [82]

Marathe, N. et al. (2013), "A treatment plant receiving waste water from multiple bulk drug manufacturers is a reservoir for highly multi-drug resistant integron-bearing bacteria.", *PloS one*, Vol. 8/10, http://dx.doi.org/10.1371/journal.pone.0077310. [26]

Markandya, A. et al. (2008), "Counting the cost of vulture decline-An appraisal of the human health and other benefits of vultures in India", *Ecological Economics*, Vol. 67/2, pp. 194-204, http://dx.doi.org/10.1016/j.ecolecon.2008.04.020. [160]

Martinez, C. et al. (2018), "In vivo study of teratogenic and anticonvulsant effects of antiepileptics drugs in zebrafish embryo and larvae", *Neurotoxicology and Teratology*, Vol. 66, pp. 17-24, http://dx.doi.org/10.1016/j.ntt.2018.01.008. [140]

Martin, M., S. Thottathil and T. Newman (2015), "Antibiotics Overuse in Animal Agriculture: A Call to Action for Health Care Providers.", *American journal of public health*, Vol. 105/12, pp. 2409-10, http://dx.doi.org/10.2105/AJPH.2015.302870. [49]

Mathias, F. et al. (2018), "Effects of low concentrations of ibuprofen on freshwater fish Rhamdia quelen", *Environmental Toxicology and Pharmacology*, Vol. 59, pp. 105-113, http://dx.doi.org/10.1016/j.etap.2018.03.008. [126]

Meinertz, J. et al. (2010), "Chronic toxicity of diphenhydramine hydrochloride and erythromycin thiocyanate to Daphnia, Daphnia magna, in a continuous exposure test system", *Bulletin of Environmental Contamination and Toxicology*, Vol. 85/5, pp. 447-451, http://dx.doi.org/10.1007/s00128-010-0117-7. [147]

Melvin, S. and F. Leusch (2016), "Removal of trace organic contaminants from domestic wastewater: A meta-analysis comparison of sewage treatment technologies", *Environment International*, Vol. 92-93, pp. 183-188, http://dx.doi.org/10.1016/J.ENVINT.2016.03.031. [31]

Metz, F. and K. Ingold (2014), "Sustainable wastewater management: Is it possible to regulate micropollution in the future by learning from the past? A policy analysis", *Sustainability (Switzerland)*, Vol. 6/4, pp. 1992-2012, http://dx.doi.org/10.3390/su6041992. [174]

Mezzelani, M. et al. (2016), "Ecotoxicological potential of non-steroidal anti-inflammatory drugs (NSAIDs) in marine organisms: Bioavailability, biomarkers and natural occurrence in Mytilus galloprovincialis", *Marine Environmental Research*, Vol. 121, pp. 31-39, http://dx.doi.org/10.1016/j.marenvres.2016.03.005. [128]

Mezzelani, M. et al. (2016), "Ecotoxicological potential of non-steroidal anti-inflammatory drugs (NSAIDs) in marine organisms: Bioavailability, biomarkers and natural occurrence in Mytilus galloprovincialis", *Marine Environmental Research*, Vol. 121, pp. 31-39, http://dx.doi.org/10.1016/J.MARENVRES.2016.03.005. [79]

Michael, I. et al. (2013), "Urban wastewater treatment plants as hotspots for the release of antibiotics in the environment: A review", *Water Research*, Vol. 47/3, pp. 957-995, http://dx.doi.org/10.1016/j.watres.2012.11.027. [13]

Miller, T. et al. (2018), "A review of the pharmaceutical exposome in aquatic fauna", *Environmental Pollution*, http://dx.doi.org/10.1016/j.envpol.2018.04.012. [180]

MistraPharma (2016), *Identification and Reduction of Environmental Risks Caused by Human Pharmaceuticals. MistraPharma Research 2008–2015. Final report.* [171]

Mompelat, S., B. Le Bot and O. Thomas (2009), "Occurrence and fate of pharmaceutical products and by-products, from resource to drinking water", *Environment International*, Vol. 35/5, pp. 803-814, http://dx.doi.org/10.1016/j.envint.2008.10.008. [68]

Monteiro, S. and A. Boxall (2010), *Occurrence and Fate of Human Pharmaceuticals in the Environment*, http://dx.doi.org/10.1007/978-1-4419-1157-5_2. [10]

Moore, S. et al. (2016), "Endogenous Estrogens, Estrogen Metabolites, and Breast Cancer Risk in Postmenopausal Chinese Women", *Journal of the National Cancer Institute*, Vol. 108/10, http://dx.doi.org/10.1093/jnci/djw103. [122]

Mullard, A. (2019), *2018 FDA drug approvals*, NLM (Medline), http://dx.doi.org/10.1038/d41573-019-00014-x. [4]

Näslund, J. et al. (2017), "Diclofenac affects kidney histology in the three-spined stickleback (Gasterosteus aculeatus) at low Mg/L concentrations", *Aquatic Toxicology*, Vol. 189, pp. 87-96, http://dx.doi.org/10.1016/j.aquatox.2017.05.017. [125]

Neale, P., F. Leusch and B. Escher (2017), "Applying mixture toxicity modelling to predict bacterial bioluminescence inhibition by non-specifically acting pharmaceuticals and specifically acting antibiotics", *Chemosphere*, http://dx.doi.org/10.1016/j.chemosphere.2017.01.018. [197]

Nelles, J., W. Hu and G. Prins (2011), "Estrogen action and prostate cancer", *Expert Review of Endocrinology and Metabolism*, Vol. 6/3, pp. 437-451, http://dx.doi.org/10.1586/eem.11.20. [123]

Niemuth, N. et al. (2015), "Metformin exposure at environmentally relevant concentrations causes potential endocrine disruption in adult male fish", *Environmental Toxicology and Chemistry*, Vol. 34/2, pp. 291-296, http://dx.doi.org/10.1002/etc.2793. [137]

NOAH (2018), *Industry Facts and Figures: National Office of Animal Health*, https://www.noah.co.uk/about/industry-facts-and-figures/ (accessed on 28 December 2018). [89]

Oaks, J. et al. (2004), "Diclofenac residues as the cause of vulture population decline in Pakistan", *Nature*, Vol. 427/6975, pp. 630-633, http://dx.doi.org/10.1038/nature02317. [109]

OECD (2018), *Stemming the Superbug Tide: Just A Few Dollars More*, OECD Health Policy Studies, OECD Publishing, Paris, https://dx.doi.org/10.1787/9789264307599-en. [166]

OECD/FAO (2018), *OECD-FAO Agricultural Outlook 2018-2027*, OECD Publishing, Paris/FAO, Rome, https://dx.doi.org/10.1787/agr_outlook-2018-en. [88]

Padhye, L. et al. (2014), "Year-long evaluation on the occurrence and fate ofpharmaceuticals, personal care products, andendocrine disrupting chemicals in an urban drinking water treatment plant", *Water Research*, Vol. 51, pp. 266-276, http://dx.doi.org/10.1016/j.watres.2013.10.070. [60]

Pal, A. et al. (2010), "Impacts of emerging organic contaminants on freshwater resources: Review of recent occurrences, sources, fate and effects", *Science of The Total Environment*, Vol. 408/24, pp. 6062-6069, http://dx.doi.org/10.1016/J.SCITOTENV.2010.09.026. [51]

Pinckney, J. et al. (2013), "Sublethal effects of the antibiotic tylosin on estuarine benthic microalgal communities", *Marine Pollution Bulletin*, Vol. 68/1-2, pp. 8-12, http://dx.doi.org/10.1016/J.MARPOLBUL.2013.01.006. [75]

Porsbring, T. et al. (2009), "Toxicity of the pharmaceutical clotrimazole to marine microalgal communities", *Aquatic Toxicology*, Vol. 91/3, pp. 203-211, http://dx.doi.org/10.1016/J.AQUATOX.2008.11.003. [142]

Prakash, V. et al. (2019), "Recent changes in populations of Critically Endangered Gyps vultures in India", *Bird Conservation International*, Vol. 29/1, pp. 55-70, http://dx.doi.org/10.1017/S0959270917000545. [161]

Product Stewardship Council (2018), *Webinar | Global Best Practices for Drug Take-Back Programs - Product Stewardship Institute (PSI)*, https://www.productstewardship.us/page/20180607_GBPFDTBP (accessed on 23 July 2018). [42]

Puckowski, A. et al. (2016), "Bioaccumulation and analytics of pharmaceutical residues in the environment: A review", *Journal of Pharmaceutical and Biomedical Analysis*, Vol. 127, pp. 232-255, http://dx.doi.org/10.1016/j.jpba.2016.02.049. [81]

Redshaw, C. et al. (2013), "Potential Changes in Disease Patterns and Pharmaceutical Use in Response to Climate Change", *Journal of Toxicology and Environmental Health, Part B*, Vol. 16/5, pp. 285-320, http://dx.doi.org/10.1080/10937404.2013.802265. [92]

Review on Antimicrobial Resistance (2015), *ANTIMICROBIALS IN AGRICULTURE AND THE ENVIRONMENT: REDUCING UNNECESSARY USE AND WASTE*. [48]

Rico, A. et al. (2012), "Use of chemicals and biological products in Asian aquaculture and their potential environmental risks: A critical review", *Reviews in Aquaculture*, Vol. 4/2, pp. 75-93, http://dx.doi.org/10.1111/j.1753-5131.2012.01062.x. [47]

Roose-Amsaleg, C. and A. Laverman (2016), "Do antibiotics have environmental side-effects? Impact of synthetic antibiotics on biogeochemical processes", *Environmental Science and Pollution Research*, Vol. 23/5, pp. 4000-4012, http://dx.doi.org/10.1007/s11356-015-4943-3. [130]

Runnalls, T. et al. (2015), "From single chemicals to mixtures-Reproductive effects of levonorgestrel and ethinylestradiol on the fathead minnow", *Aquatic Toxicology*, Vol. 169, pp. 152-167, http://dx.doi.org/10.1016/j.aquatox.2015.10.009. [195]

Saad, W. et al. (2017), "Drug product immobilization in recycled polyethylene/polypropylene reclaimed from municipal solid waste: experimental and numerical assessment", *Environmental Technology (United Kingdom)*, Vol. 38/23, pp. 3064-3073, http://dx.doi.org/10.1080/09593330.2017.1288271. [40]

Säfholm, M. et al. (2015), "Mixture effects of levonorgestrel and ethinylestradiol: Estrogenic biomarkers and hormone receptor mRNA expression during sexual programming", *Aquatic Toxicology*, Vol. 161, pp. 146-153, http://dx.doi.org/10.1016/j.aquatox.2015.02.004. [190]

Säfholm, M. et al. (2012), "Disrupted Oogenesis in the Frog Xenopus tropicalis after Exposure to Environmental Progestin Concentrations", *Biology of Reproduction*, Vol. 86/4, http://dx.doi.org/10.1095/biolreprod.111.097378. [189]

SAICM (2015), *Nomination for new emerging policy issue: environmentally persistent pharmaceutical pollutants*, http://www.saicm.org/Portals/12/documents/meetings/ICCM4/doc/K1502367%20SAICM-ICCM4-7-e.pdf (accessed on 23 January 2019). [175]

Santos, L. et al. (2010), "Ecotoxicological aspects related to the presence of pharmaceuticals in the aquatic environment", *Journal of Hazardous Materials*, Vol. 175/1-3, pp. 45-95, http://dx.doi.org/10.1016/J.JHAZMAT.2009.10.100. [108]

Schmidt, S., J. Winter and C. Gallert (2012), "Long-Term Effects of Antibiotics on the Elimination of Chemical Oxygen Demand, Nitrification, and Viable Bacteria in Laboratory-Scale Wastewater Treatment Plants", *Archives of Environmental Contamination and Toxicology*, Vol. 63/3, pp. 354-364, http://dx.doi.org/10.1007/s00244-012-9773-4. [164]

Schultz, M. et al. (2011), "Selective uptake and biological consequences of environmentally relevant antidepressant pharmaceutical exposures on male fathead minnows", *Aquatic Toxicology*, Vol. 104/1-2, pp. 38-47, http://dx.doi.org/10.1016/j.aquatox.2011.03.011. [153]

Schwarzenbach, R. et al. (2006), "The challenge of micropollutants in aquatic systems", *Science*, Vol. 313/5790, pp. 1072-1077, http://dx.doi.org/10.1126/science.1127291. [173]

Scott, T. et al. (2018), "Pharmaceutical manufacturing facility discharges can substantially increase the pharmaceutical load to U.S. wastewaters", *Science of the Total Environment*, Vol. 636, pp. 69-79, http://dx.doi.org/10.1016/j.scitotenv.2018.04.160. [27]

Segura, P. et al. (2015), "Global occurrence of anti-infectives in contaminated surface waters: Impact of income inequality between countries", *Environment International*, Vol. 80, pp. 89-97, http://dx.doi.org/10.1016/j.envint.2015.04.001. [83]

Silva, L. et al. (2017), "SSRIs antidepressants in marine mussels from Atlantic coastal areas and human risk assessment", *Science of the Total Environment*, Vol. 603-604, pp. 118-125, http://dx.doi.org/10.1016/j.scitotenv.2017.06.076. [97]

Singer, A. et al. (2014), "Intra- and inter-pandemic variations of antiviral, antibiotics and decongestants in wastewater treatment plants and receiving rivers", *PLoS ONE*, Vol. 9/9, http://dx.doi.org/10.1371/journal.pone.0108621. [54]

Snyder, S. et al. (2008), *Toxicological Relevance of EDCs and Pharmaceuticals in Drinking Water*, AWWA Research Foundation, Denver, CO. [105]

Solé, M. et al. (2010), "Effects on feeding rate and biomarker responses of marine mussels experimentally exposed to propranolol and acetaminophen", *Analytical and Bioanalytical Chemistry*, Vol. 396/2, pp. 649-656, http://dx.doi.org/10.1007/s00216-009-3182-1. [76]

Stoker, T., E. Gibson and L. Zorrilla (2010), "Triclosan exposure modulates estrogen-dependent responses in the female wistar rat", *Toxicological Sciences*, Vol. 117/1, pp. 45-53, http://dx.doi.org/10.1093/toxsci/kfq180. [143]

Sun, Q. et al. (2014), "Seasonal variation in the occurrence and removal of pharmaceuticals and personal care products in a wastewater treatment plant in Xiamen, China", *Journal of Hazardous Materials*, Vol. 277, pp. 69-75, http://dx.doi.org/10.1016/j.jhazmat.2013.11.056. [55]

Svensson, J. et al. (2014), "Environmental concentrations of an androgenic progestin disrupts the seasonal breeding cycle in male three-spined stickleback (Gasterosteus aculeatus)", *Aquatic Toxicology*, Vol. 147, pp. 84-91, http://dx.doi.org/10.1016/j.aquatox.2013.12.013. [117]

Svensson, J. et al. (2013), "The synthetic progestin levonorgestrel is a potent androgen in the three-spined stickleback (Gasterosteus aculeatus)", *Environmental Science and Technology*, Vol. 47/4, pp. 2043-2051, http://dx.doi.org/10.1021/es304305k. [118]

Svensson, J. et al. (2016), "Developmental exposure to progestins causes male bias and precocious puberty in zebrafish (Danio rerio)", *Aquatic Toxicology*, Vol. 177, pp. 316-323, http://dx.doi.org/10.1016/j.aquatox.2016.06.010. [119]

Thrupp, T. et al. (2018), "The consequences of exposure to mixtures of chemicals: Something from 'nothing' and 'a lot from a little' when fish are exposed to steroid hormones", *Science of the Total Environment*, Vol. 619-620, pp. 1482-1492, http://dx.doi.org/10.1016/j.scitotenv.2017.11.081. [196]

Tong, A., B. Peake and R. Braund (2011), "Disposal practices for unused medications around the world", *Environment International*, Vol. 37/1, pp. 292-298, http://dx.doi.org/10.1016/j.envint.2010.10.002. [39]

Tran, N., M. Reinhard and K. Gin (2018), *Occurrence and fate of emerging contaminants in municipal wastewater treatment plants from different geographical regions-a review*, http://dx.doi.org/10.1016/j.watres.2017.12.029. [33]

UN Environment (2019), *Global Chemicals Outlook II: From legacies to innovative solutions*, United Nations Environment Programme, https://wedocs.unep.org/bitstream/handle/20.500.11822/27651/GCOII_synth.pdf?sequence=1&isAllowed=y (accessed on 27 June 2019). [85]

Van Boeckel, T. (2017), *A Global Plan To Cut Antimicrobial Use In Animals*, https://cddep.org/blog/posts/global-plan-cut-antimicrobial-use-animals/ (accessed on 11 September 2018). [18]

Van Boeckel, T. et al. (2015), "Global trends in antimicrobial use in food animals.", *Proceedings of the National Academy of Sciences of the United States of America*, Vol. 112/18, pp. 5649-54, http://dx.doi.org/10.1073/pnas.1503141112. [90]

Vatovec, C. et al. (2016), "Investigating dynamic sources of pharmaceuticals: Demographic and seasonal use are more important than down-the-drain disposal in wastewater effluent in a University City setting", *Science of the Total Environment*, Vol. 572, pp. 906-914, http://dx.doi.org/10.1016/j.scitotenv.2016.07.199. [58]

Verlicchi, P., M. Al Aukidy and E. Zambello (2012), "Occurrence of pharmaceutical compounds in urban wastewater: Removal, mass load and environmental risk after a secondary treatment-A review", *Science of the Total Environment*, Vol. 429, pp. 123-155, http://dx.doi.org/10.1016/j.scitotenv.2012.04.028. [15]

Verlicchi, P. et al. (2010), "Hospital effluents as a source of emerging pollutants: An overview of micropollutants and sustainable treatment options", *Journal of Hydrology*, Vol. 389/3-4, pp. 416-428, http://dx.doi.org/10.1016/j.jhydrol.2010.06.005. [16]

Vestel, J. et al. (2016), "Use of acute and chronic ecotoxicity data in environmental risk assessment of pharmaceuticals", *Environmental Toxicology and Chemistry*, Vol. 35/5, pp. 1201-1212, http://dx.doi.org/10.1002/etc.3260. [141]

Vulliet, E. and C. Cren-Olivé (2011), "Screening of pharmaceuticals and hormones at the regional scale, in surface and groundwaters intended to human consumption", *Environmental Pollution*, Vol. 159/10, pp. 2929-2934, http://dx.doi.org/10.1016/j.envpol.2011.04.033. [70]

Watkinson, A. et al. (2009), "The occurrence of antibiotics in an urban watershed: From wastewater to drinking water", *Science of The Total Environment*, Vol. 407/8, pp. 2711-2723, http://dx.doi.org/10.1016/J.SCITOTENV.2008.11.059. [52]

Weber, F. et al. (2014), *Pharmaceuticals in the environment - The global perspective: Occurrence, effects, and potential cooperative action under SAICM*, German Federal Environmental Agency. [3]

WHO (2017), *Chemical mixtures in source water and drinking-water*, World Health Organisation, Geneva. [201]

WHO (2012), *Pharmaceuticals in Drinking Water*, World Health Organisation, Geneva. [2]

WHO (2012), *State of the Science of Endocrine Disrupting Chemicals 2012 Summary for Decision-Makers INTER-ORGANIZATION PROGRAMME FOR THE SOUND MANAGEMENT OF CHEMICALS*, World Health Organization. [115]

Xia, L., L. Zheng and J. Zhou (2017), "Effects of ibuprofen, diclofenac and paracetamol on hatch and motor behavior in developing zebrafish (Danio rerio)", *Chemosphere*, Vol. 182, pp. 416-425, http://dx.doi.org/10.1016/j.chemosphere.2017.05.054. [127]

Yamagishi, T., Y. Horie and N. Tatarazako (2017), "Synergism between macrolide antibiotics and the azole fungicide ketoconazole in growth inhibition testing of the green alga Pseudokirchneriella subcapitata", *Chemosphere*, Vol. 174, pp. 1-7, http://dx.doi.org/10.1016/j.chemosphere.2017.01.071. [198]

Yang, Y. et al. (2017), *Occurrences and removal of pharmaceuticals and personal care products (PPCPs) in drinking water and water/sewage treatment plants: A review*, http://dx.doi.org/10.1016/j.scitotenv.2017.04.102. [32]

Yu, Y., L. Wu and A. Chang (2013), "Seasonal variation of endocrine disrupting compounds, pharmaceuticals and personal care products in wastewater treatment plants", *Science of the Total Environment*, Vol. 442, pp. 310-316, http://dx.doi.org/10.1016/j.scitotenv.2012.10.001. [56]

Zounková, R. et al. (2007), "Ecotoxicity and genotoxicity assessment of cytostatic pharmaceuticals", *Environmental Toxicology and Chemistry*, Vol. 26/10, p. 2208, http://dx.doi.org/10.1897/07-137R.1. [134]

Notes

1 The total volume of medically important antibiotics sold for use in food animals based on 2014 FDA data was 9,479,339 kg. The total volume of antibiotics sold for use in humans based on IMS Health calculations was 3,494,673 kg. The figures are rounded to 73% used in animals and 27% used in humans.

2 Global meat production is projected to be 15% higher in 2027 relative to a base period of 2015-17. Over the same time period, global fish production is expected to increase by 13.4%. World growth in fish production will be completely founded upon the continued growth in aquaculture output; capture fisheries production is expected to fall (OECD/FAO, 2018[88]).

3 For example, information is well established and increasing for the following pharmaceuticals: diclofenac, paracetamol, ibuprofen, carbamazepine, naproxen, atenolol, ethinyloestradiol, aspirin, fluoxetine, propranolol, metoprolol and sulfamethoxazole.

4 In many developing nations, antibiotics may be purchased over-the-counter, without a doctor's prescription.

5 Pharmacodynamics is the study of how a drug affects an organism. Pharmacokinetics is the study of how an organism affects a drug. Both together influence dosing, benefit, and adverse effects.

2 Opportunities to build a policy-relevant knowledge base

Improving knowledge on pharmaceuticals in water, and their effects and risks on human health and the environment, is an important foundation on which pharmaceutical authorisation, environmental risk assessments and water quality policies can be built. This chapter inventories and assesses the strengths and weaknesses of various innovative monitoring and modelling approaches to assess the impacts and risks of pharmaceuticals in water. The chapter takes stock of country and international initiatives to improve the knowledge base and highlights the need for data sharing and institutional coordination.

2.1. Key messages

Numerous active pharmaceutical ingredients (APIs) have been detected in surface waters as a result of advancements in analytical technologies. Certain pharmaceuticals are being monitored in surface waters of OECD countries according to watch-lists, although the vast majority of them remain unmonitored. Furthermore, many pharmaceuticals in use have not been assessed for their environmental impact, and the environment is not considered in the risk-benefit analysis when authorising new medicines for human use.

Knowledge gaps persist on the sources, transport and fate of APIs (including transformation products and metabolites) in water bodies, and their toxicological and chronic impacts (including additive effects of mixtures) to a multitude of targets in organisms, populations, and communities, over short and long-terms. Member countries are looking for innovative ways to maximise the benefits of existing knowledge, and develop new tools and decision frameworks that are cost- and time-effective.

Without adequate knowledge of the potential hazards APIs may pose, assessing their environmental risk is challenging. As it would be impractical (technically and economically) to commence analytical monitoring *en masse* for all APIs in the environment, more integrated and holistic approaches to environmental monitoring are needed, in combination with modelling, prioritisation methods and data sharing initiatives.

In order to address knowledge gaps, perform robust environmental risk assessments and design informed policy responses, it is necessary to establish standardisation and guidance on the best practices in analytical methods and risk assessment, data quality and exchange, and prioritisation of pharmaceuticals and susceptible water bodies. Country and international initiatives are crucial to improve the knowledge base, and the exchange and review of data.

Lastly, it is important to note that improving knowledge is not a pollution reduction measure in itself. Governments should take advantage of alternative innovative monitoring and modelling technologies to simplify the prioritisation and identification of APIs of concern, and improve the speed and quality of risk assessments and cost benefit analysis, but they should not wait for exact science before taking action.

2.2. Environmental risk assessment and authorisation of pharmaceuticals

2.2.1. Introduction

One of the key legislative factors influencing the presence of the pharmaceuticals in the environment is the current framework for Environmental Risk Assessment (ERA), which is a part of the Market Authorisation process of new pharmaceuticals. The objective of an ERA is to determine the potential adverse effects that pharmaceuticals pose to ecological health, and is a combined evaluation of hazards and exposure. Toxicity is determined through toxicity testing (*in vivo*), which involves an assessment of the harm to aquatic organisms. To assess whether exposure levels are safe, scientists calculate the ratio of the predicted environmental concentration (PEC; based on distributions) to the predicted no-effect concentration (PNEC; the predicted level below which there are no negative consequences for the ecosystem, or an ecological safety threshold). Risk is identified when substances are present in water at higher levels than would be safe for the environment (PEC:PNEC ratios greater than one) (EMA, 2006[1]). Of specific concern are substances which are persistent, bioaccumulative and toxic (PBT), or very persistent and very bioaccumulative (vPvB) (see section 1.4.2), and substances with endocrine disruptive (where not intended), mutagenic or carcinogenic properties.

Most (88%) of the pharmaceuticals targeting human proteins do not have comprehensive environmental toxicity data, or they are not publically available. This highlights the need for both intelligent approaches to prioritise legacy human drugs for a tailored environmental risk assessment and a transparent database that captures environmental data (Gunnarsson et al., 2019[2]).

Due to increased awareness regarding pharmaceuticals in the environment, pharmaceutical producers and regulatory bodies have started to assess potential environmental risks when designing and authorising new pharmaceuticals. The following sections outline current ERA practice in countries regarding human and veterinary pharmaceuticals, and how they are used in decision-making when new pharmaceuticals are being authorised.

2.2.2. Environmental risk assessment of new pharmaceuticals in practice

Within the EU, since 2006, an ERA is required for all new marketing authorisations of human pharmaceuticals according to Article 8(3) of Directive 2001/83/EC (Box 2.1). Pharmaceuticals entered on the market before 2006 do not require an ERA in a retrospective manner, unless a line extension (new version or an enhancement of an existing product) is requested. An important point to note is that even if a risk is identified in the ERA, it is not considered in the benefit-risk assessment of authorisation, and therefore cannot be used as grounds for refusal in marketing. However, on a case-by-case basis, specific arrangements to limit impact on the environment can be considered including, for example, specific labelling (EMA, 2006[1]).

Box 2.1. Environmental Risk Assessment in the authorisation process of new human pharmaceuticals (post 2006), EU

Realising that pharmaceuticals could pose an environmental risk, the European Medicines Agency developed guidance for ERAs in 2006. During the authorisation process, the applicant is required to evaluate environmental impact, in terms of exposure and effects, and submit the assessment of the potential risk. The assessment is a two-phase procedure (see figure below) initiated with Phase I, where the predicted environmental concentration (PEC) for surface water is calculated and the distribution is measured. If the PEC is equal to, or above, a trigger value of 0.01 µg/L, a second phase of analysis is carried out. Substances that are known to affect the reproduction of organisms at low concentrations, or have the potential to bioaccumulate, should also enter Phase II.

Phase II is a two-tiered (A and B) environmental fate and effect analysis. In this phase, the properties of the substance (persistence, bioaccumulation and toxicity) are investigated. In Phase II A, PNECs are calculated for surface water, groundwater and microorganisms based on a standard long-term toxicity test on fish, daphnia and algae (e.g. according to OECD Test Guidelines 201, 211 and 210). When the results suggest potential risk, Phase II is extended for further evaluation on the fate of the substance and/or its metabolites in the aquatic environment (Phase II B) (EMA, 2006[1]).

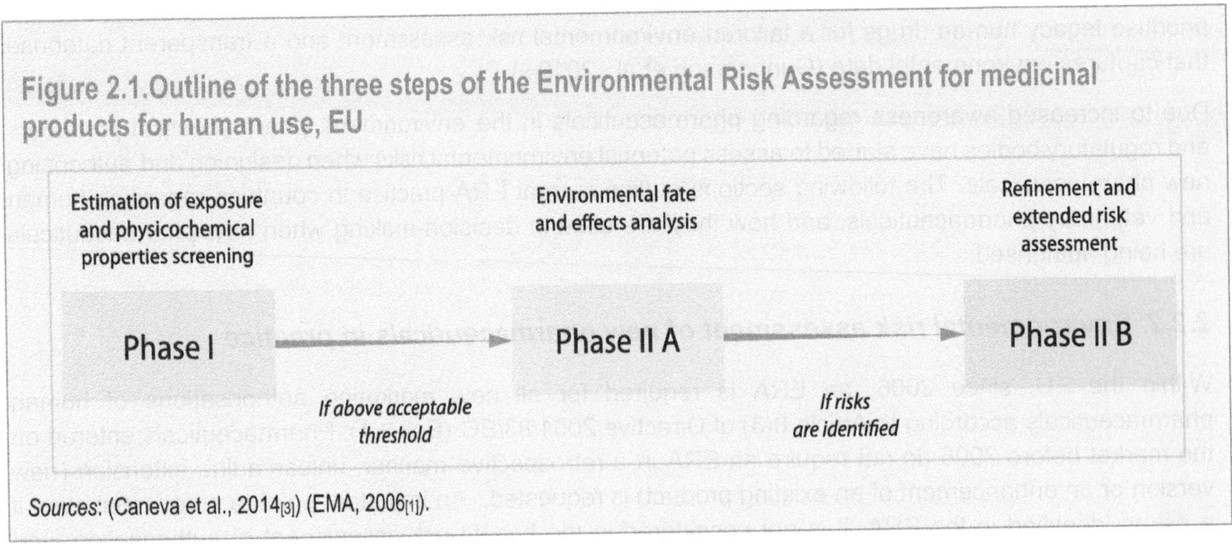

Figure 2.1.Outline of the three steps of the Environmental Risk Assessment for medicinal products for human use, EU

Sources: (Caneva et al., 2014[3]) (EMA, 2006[1]).

Similar to the EU, ERAs are required by the U.S. Food and Drug Administration (FDA) for new (post 2008) drug applications (with some exceptions). New FDA (2016[4]) guidance supplements the Environmental Assessment Guidance issued in (1998[5]) by addressing specific considerations for drugs that have potential oestrogenic, androgenic, or thyroid hormone pathway activity in the environment, and the conditions under which the sponsor should submit an environmental assessment or may apply for a claim of categorical exclusion. However, also like the EU, the results are not considered as part of the benefit-risk assessment of authorisation.

Canada's New Substances Program is responsible for administering the *New Substances Notification Regulations (Chemicals and Polymers)* and the *New Substances Notification Regulations (Organisms)* of the *Canadian Environmental Protection Act, 1999* (CEPA). Collectively known as the 'Regulations', they are an integral part of the federal government's national pollution prevention strategy. As part of the "cradle to grave" management approach for toxic substances laid out in CEPA, the Regulations were created to ensure that no new substances (chemicals, polymers or animate products of biotechnology) are introduced into the Canadian marketplace before an assessment of whether they are potentially toxic has been completed, and any appropriate or required control measures have been taken. In September 2001, substances in products regulated by the *Food & Drugs Act* became subject to the Regulations. This includes substances used in pharmaceuticals, veterinary drugs, biologics (including genetic therapies), medical devices, cosmetics and personal care products, food additives, novel foods and natural health products. Under the Regulations, a New Substances Notification package, containing all information prescribed in the Regulations, may be required before a new substance can be imported into or manufactured in Canada. The type of information required and the timing of the notification will depend on such factors as the type of substance, the quantity that will be imported or manufactured, the intended use of the substance and the circumstances associated with its introduction. When a potential risk to human health or the environment is identified for a new substance, CEPA empowers the Government of Canada to develop risk management measures prior to or during the earliest stages of its introduction into Canada. This ability to act early makes the New Substances Program a unique and essential component of the federal management of toxic substances.

The International Cooperation on Harmonisation of Technical Requirements for Registration of Veterinary Medicinal Products (VICH), launched in 1996, provides a basis for harmonising technical requirements for veterinary product registration in the EU, U.S. and Japan. Observers of the VICH also include Australia, New Zealand, Canada and South Africa. The VICH platform stipulates guidelines on how environmental impact assessments (EIA) should be performed (Box 2.2), but does not constitute a role to provide guidance to establish regulatory systems or regulations for marketing authorisation of veterinary pharmaceuticals (VICH, 2018[6]).

> **Box 2.2. The International Cooperation on Harmonisation of Technical Requirements for Registration of Veterinary Medicinal Products: Guidance on EIA**
>
> The International Cooperation on Harmonisation of Technical Requirements for Registration of Veterinary Medicinal Products (VICH) provides guidance on carrying out environmental impact assessments of veterinary pharmaceuticals. The guidance builds on a two-phase approach, similar to the one used for human pharmaceuticals in the EU (see Box 2.1):
>
> - Phase I builds on a series of questions that the applicant must answer and predicted environmental concentration (PEC) calculations, which decides if the pharmaceutical in question should enter Phase II investigation. The trigger values (i.e. calculated PEC) for Phase II are 100 µg/kg and 1 µg/L for soil and indirect aquatic exposure, respectively. Certain therapeutic groups automatically enter Phase II regardless of the PEC calculation, e.g. ecto- and endoparasiticides intended for use in livestock (VICH, 2000[7]).
>
> - Phase II requires an assessment of the environmental fate and effects of the pharmaceutical, and includes studies of: (1) physico-chemical properties (e.g. water solubility, octanol/water partitioning); (2) binding to soils; (3) biodegradation in soil or aquatic test systems; and (4) acute effects of the drug residue on select aquatic and terrestrial species (Tier A). The approach is not fully harmonised in the VICH regions due to differences in animal husbandry and land-use practices. Results indicating a risk can require further study (Tier B), for example, when substances have a logKow distribution greater than 4, which signals potential for bioaccumulation (EMA, 2004[8]).
>
> *Sources*: (EMA, 2004[8]) (VICH, 2000[7]).

Within the EU, in contrast to human pharmaceuticals, the marketing authorisation of veterinary medicines does require the environmental risks to be included in the benefit-risk assessment (Directive 2001/82/EC, as amended by Directive 2004/28/EC), i.e. the benefits are weighed up against the environmental risks before putting them on the market (EMA, 2009[9]). However, the current guidelines on benefit-risk are not clear on how trade-offs should be assessed (Chapman et al., 2017[10]). Generally, veterinary pharmaceuticals are authorised if the benefits are thought to outweigh the environmental risks or because no alternative treatment is available (UBA, 2018[11]). In 2016, a guideline was adopted within the EU for hazard-based assessment of PBT and vPvB substances used in veterinary medicinal products. So far, all potential PBTs identified belong to the therapeutic group of parasiticides (EMA, 2017[12]).

Since the introduction of pharmaceutical regulations in EU, the German Environment Agency (UBA) has been evaluating ERAs for human and veterinary medicines before they are marketed. They estimate that 10% of pharmaceutical products indicate a potential environmental risk (Küster and Adler, 2014[13]). Of greatest concern are hormones, antibiotics, analgesics, antidepressants and anticancer pharmaceuticals used for human health, and hormones, antibiotics and parasiticides used as veterinary pharmaceuticals (Küster and Adler, 2014[13]).

2.2.3. Concerns with current environment risk assessment of pharmaceuticals

Several researchers and OECD member countries stress the limitations of current practices for ERAs for the authorisation of human and veterinary pharmaceuticals:

- ERAs are not considered in the risk-benefit analysis of marketing authorisation for human pharmaceuticals; current approval for human pharmaceuticals is based on safety, efficacy and quality. This issue has been raised by several scientists (e.g. (Ågerstrand et al., 2015[14]) (Küster and Adler, 2014[13])) as well as within government national action plans related to pharmaceuticals in the environment. Germany for instance, stresses the need to incorporate the environment as a factor in the benefit-risk balance (similar to the framework for veterinary pharmaceuticals). Sweden has also pushed for the inclusion of environmental risk in the benefit-risk analysis of human pharmaceuticals, as a driver for pharmaceutical companies to take greater responsibility and implement environmental risk mitigation measures when necessary.

- ERAs are not required retrospectively for already-authorised pharmaceuticals (pre 2006 in the EU and pre-2008 in the US) that may be of high environmental risk, and there is no formal mechanism to review assessments as information on the risks improve. Potential environmental risks of human and veterinary pharmaceuticals have been identified for different pharmaceutical classes such as parasiticides, analgesics, antidepressants, antibiotics, hormones and anticancer medicines (see section 1.4). Within the EU, there are about 2000 veterinary pharmaceutical products on the market, most of which have not been fully tested for their toxicity (Kools et al., 2008[15]). Likewise, most (88%) human pharmaceuticals do not have comprehensive environmental toxicity data, or they are not publically available (Gunnarsson et al., 2019[2]).

- In the current ERA process, some pharmaceuticals do not reach the trigger value for a thorough assessment (i.e. may stop after the first phase of an ERA). For example, endocrine active compounds have to be assessed, but others, such as cytotoxics, are not assessed despite their potential high hazard (Kümmerer et al., 2016[16]). ERAs also lack consideration of antimicrobial resistance; a more diverse selection of bacteria in ERA would increase protectiveness of ERA (Ågerstrand et al., 2015[14]; Le Page et al., 2017[17]; Le Page et al., 2019[18]).

- ERAs for pharmaceuticals are product-based, and not assessed per API used within the product. This may lead to different conclusions about the risk for the same API used in different pharmaceutical products. It also duplicates the laboratory work, and the animals being used for testing (Ågerstrand et al., 2015[14]). Moreover, this complicates the search for information in a systematic way.

- There is no uniform regulation for ERA of mixture and additive effects. There is a lack of peer-reviewed toxicity data for commonly identified chemical mixtures which would enable assessment of the hazard as a whole rather than being based on individual components (WHO, 2017[19]). For example, Cleuvers (2008[20]) found that toxicity of a mixture of non-steroidal anti-inflammatory drugs against Daphnia was considerably higher even at concentrations in which the single substances showed no, or only very slight, effects. Reproduction was decreased by 100% at concentrations where no effects on survival could be observed, which means that this destructive effect on the Daphnia population would be totally overlooked by an acute test using the same concentrations. Section 1.4.2 outlines how mixtures of pharmaceuticals, and with other contaminants in the environment, can possess a joint toxicity greater than the toxicity of individual substances. Some efforts have been made within the EU to establish guidance for chemical mixtures, for example, the State of Art Report on Mixture Toxicity (Kortenkamp Andreas and Faust, 2009[21]) and the Communication from the European Commission on Combination effects of Chemicals[1] (Godoy and Kummrow, 2017[22]).

- There is no uniform regulation for ERA of metabolites and transformation products, and multiple routes of exposure in OECD countries. The ecotoxicity of some metabolites and transformation products can have higher additive and toxicity effects than that of their parent pharmaceutical compounds (Godoy and Kummrow, 2017[22]). Furthermore ecosystems and humans may be continuously exposed via a number of pathways of low-dose mixtures that can have additive effects (Backhaus, 2014[23]).

- When, as identified by ERA, human or veterinary pharmaceuticals pose a risk to the environment, risk mitigation measures can be recommended. However, compliance with risk mitigation measures is not enforced and is essentially voluntary (BIO Intelligence Service, 2013[24]).

- ERAs and pharmaceutical authorisation is resource-intensive, although it is less costly than pesticide or biocide registration. In 2016, the average number of new active pharmaceutical ingredients registered by OECD governments was 34, and the cost of non-clinical testing of such substances is likely to be several million euros (OECD, 2019[25]).

- Lack of involvement of different stakeholders and data sharing. The framework for ERAs for pharmaceuticals is unlike other regulations of chemicals (e.g. the EU Registration, Evaluation, Authorisation and Restriction of Chemicals (REACH) regulation) where data sharing is encouraged and stakeholders are invited to comment on draft opinions of the authorisation process and ERAs (Ågerstrand et al., 2017[26]). Since ERA information cannot be cross-referenced, data cannot be reused from one dossier to another, even if the concerned medicinal products contain the same API.

Further thinking on the ERA framework is needed to address the above challenges, including the potential for grouping risks based on the properties of pharmaceuticals (e.g. toxicity, mobility, persistence) and the receiving water (e.g. surface water, groundwater, drinking water) in order to predict, identify and mitigate future emerging environmentally persistent pharmaceutical pollutants. There is a need to better understand human and environmental exposures, through the use of both investigative monitoring and modelling.

Although the efficiency, efficacy and cost of undertaking risk assessments, as well as cost of the control of potential mixtures in relation to the health benefits of pharmaceuticals need to be carefully considered, a number of useful tools and models have been, and continue to be, developed to carry out ERA of chemical mixtures and overcome some of the challenges listed above. For example, OECD guidance on Considerations for Assessing the Risks of Combined Exposure to Multiple Chemicals (OECD, 2018[27]) can help. The World Health Organisation (2017[19]) also provides guidance on the risk assessment and management of chemical mixtures (Box 2.3).

More systematic studies will help to further the understanding of the transport, occurrence, exposure pathways and fate of pharmaceuticals in the environment, as well as susceptible species and relevant effects endpoints. Standardisation of protocols for sampling and analysing pharmaceuticals would help to facilitate the comparison of data (WHO, 2012[28]; WHO, 2017[19]). The following sections of this chapter look at existing frameworks for monitoring pharmaceuticals in the environment, and documents new research developments in monitoring technologies and approaches.

Box 2.3. UN World Health Organisation guidance on managing the risks of chemical mixtures

The WHO proposes a number of questions that can assist in making decisions about whether to treat a group of substances as a mixture for risk assessment, and what management approaches may be considered.

- Do the chemicals always occur as a mixture and, if not, how frequently do they occur together and under what circumstances?
- Does the proportion of substances vary and is there a small number that usually dominate?
- Are the substances of similar water solubility?
- Can one or two substances act as a surrogate for the others (for both risk assessment and management)?
- How stable is the mixture (i.e. is it always similar)?
- Can the components of the mixture be measured by the same method?
- How readily will components of the mixture be removed in the available wastewater and drinking-water treatment?
- Are there other upstream interventions that can be applied?

Source: (WHO, 2017[19]).

2.3. Existing frameworks for monitoring pharmaceuticals in water

Monitoring refers to repeated measurements of predetermined, specific endpoints over broad spatial and temporal scales aiming to assess the status and trends of water bodies (Ekman et al., 2013[29]). Many OECD countries have monitoring programmes to measure the concentrations of selected pharmaceuticals in surface water in order to determine appropriate measures to address the risk posed by these substances in the future (e.g. inclusion in systematic monitoring or the development of Environmental Quality Norms). However, they monitor only a selected number of pharmaceuticals. Groundwater and drinking water sources are less frequently monitored. In developing economies, monitoring of pharmaceuticals is often not a priority. Monitoring data is particularly under-represented in Asia, Africa and South America - the very regions of the world that are likely to have the highest consumption and release of pharmaceuticals due to high population density, limited wastewater treatment (aus der Beek et al., 2016[30]), and in some locations, pharmaceutical manufacturing. This section provides a brief summary of selected pharmaceutical monitoring programmes in OECD countries.

The EU Watch List under the Water Framework Directive requires monitoring of several pharmaceuticals (hormones and antibiotics) in surface water (Box 2.4). Research is underway to develop a groundwater watch list (Box 2.5). Korea created a candidate list of contaminants of emerging concern (CECs) for surface water monitoring in 2016 that includes eight pharmaceuticals (Acetylsalicylic acid, Sulfamethoxazole, Sulfamethazine, Sulfathiazole, Naproxen, Clarithromycin, Trimethoprim and Carbamazepine). In the U.S., the 1996 Safe Drinking Water Act (SDWA requires) that once every five years, the EPA publish a list of currently unregulated contaminants, known as the Contaminant Candidate List (CCL). In developing the CCL, the SDWA directs the EPA to consider health effects and occurrence information for unregulated contaminants and further specifies that the Agency place those contaminants on the list that present the greatest public health concern related to exposure from drinking water. As part of the CCL process, the EPA evaluates contaminants that are generally considered to be contaminants of emerging concern in drinking water, including pharmaceuticals. The SDWA also requires the EPA to monitor public water systems, every five years, for no more than 30 unregulated contaminants under the Unregulated

Contaminant Monitoring Rule (UCMR). The EPA's selection of contaminants for a particular UCMR cycle is largely based on a review of the CCL (US EPA, 2016[31]).

Box 2.4. Monitoring of pharmaceuticals in surface water, as required under the EU Water Framework Directive

Under the Water Framework Directive (WFD), the surface water Watch List is a list of potential water pollutants that are required to be monitored and reported by EU Member States to determine the risk they pose to the aquatic environment and whether EU Environmental Quality Norms (EQN) should be set for them. The list is reviewed every 2 years.

The first Watch List was published in 2015. It included 10 substances or groups of substances which included the hormones 17-Alpha-ethinylestradiol (EE2), 17-Beta-estradiol (E2) and oestrone (E1), and the pain killer diclofenac.

In 2018, the Watch List was updated to remove diclofenac as sufficiently high-quality monitoring data had been collected. Five antibiotics were added to the list: 17-alpha-ethinylestradiol (EE2), 17-beta-estradiol (E2), oestrone (E1), macrolide antibiotics (erythromycin, clarithromycin, azithromycin), amoxicillin and ciprofloxacin. The inclusion of antibiotics is consistent with the European One Health Action Plan against Antimicrobial Resistance (AMR), which supports the use of the Watch List to improve knowledge of the occurrence and spread of antimicrobials in the environment.

Source: (EC, 2018[32])

Box 2.5. EU voluntary watch list for groundwater monitoring of pharmaceuticals

A Watch List for substances in groundwater is currently being developed in accordance with EU Directive 2014/80/EU (amending the Groundwater Directive, 2006/118/EC, under the WFD umbrella) to increase the availability of monitoring data on substances posing a risk or potential risk to bodies of groundwater, for which groundwater quality standards or threshold values should be set. Contrary to the Watch List for surface water, the monitoring will be voluntary (instead of mandatory).

Twelve EU member states submitted pharmaceutical datasets for a pilot-study aiming to gather and identify evidence to derive a watch list for groundwater. In these 12 member states, a wide range of pharmaceuticals (in total, approximately 300 different substances) were monitored and detected in groundwater. The table below presents the 17 substances most frequently detected above 0.1 µg/L in all 12 member states. Carbamazepine was most widely analysed, followed by diclofenac. Paracetamol was the most frequently detected substance as a percentage of sites monitored (24% of the sites). In addition to these substances, diatrizoic acid, primidon, ibuprofen and clofibrate are prominent in the summary data.

Table 2.1. Substances being detected above 0.1 µg/L more than twice in groundwater of 12 EU Member States

Substance	Total number of sites	>LoQa	>0.1 µg/l
ethylenediaminetetraacetic acid (EDTA)	105	61	61
carbamazepine	3692	445	57
ibuprofen	1865	49	15
diatrizoic acid	1394	66	15
diclofenac	3060	34	13
gabapentin	167	19	13
metformin	787	56	12
paracetamol (acetaminophen)	1036	248	11
clofibrate	2294	46	10
primidon	628	31	10
pentoxifylline	859	9	9
acetylsalicylic acid	600	7	7
phenazone	1477	40	5
sulfamethoxazole	2153	105	4
ioxithalamic acid	329	12	4
erythromycin	1255	25	3
iopamidol	1462	37	3

a. Number of samples above limit of quantification (LoQ)
Source: (Marsland and Roy, 2016[33]).

Taking monitoring results one step further, environmental quality norms (EQNs) can be developed to define the maximum allowable concentration of a single substance in water to protect ecosystems, and drinking water and bathing resources. The EU Water Framework Directive (WFD) (2000/60/EC), among others, builds on EQNs to secure water quality.

The European Commission suggested EQNs for each of the pharmaceuticals on the surface water Watch List (Box 2.4), but they are not incorporated in EU-legislation. Under the WFD, member states are required to select additional substances of national or local concern (so called river basin-specific substances), and to define corresponding EQNs (WFD Annex VIII) (European Commission, 2017[34]). Sweden has proposed 17 pharmaceuticals for monitoring in addition to the WFD Watch List, based on PBT properties, large usage, and/or detection in fish, surface water, drinking water and sludge (MPA, 2015[35]). In addition, Sweden has incorporated EQNs for 4 pharmaceuticals (Ciprofloxacin, Diclofenac, E2 and EE2) as river basin-specific substances according to the Swedish Agency for Marine and Water Management statues HVMFS 2018:17. The U.K has transposed obligation to monitor substances on the watch-list into national legislation. The Netherlands has recently proposed EQNs for Carbamazepine, Metoprolol and Metformin (RIVM, 2014[36]). Australia has developed regulatory guidelines for water recycling including drinking water-based guidelines (ranging from 0.35 to 1050 µg/L) for antibiotics, non-steroidal anti-inflammatories, β-adrenergic blockers, estrogenic hormones, and other general pharmaceuticals (Australian Government, 2008[37]).

It has been acknowledged that the number of pharmaceuticals measured by target analysis is not sufficient to provide an exhaustive overview of water quality; target-based environmental monitoring neglects unknown but potentially substantial portions of constituents in the aquatic environment (Daughton, 2004[38]). Hence, substances that are often detected can lead to the assumption that they are the most common, when in reality they are simply the most studied (aus der Beek et al., 2016[30]). Conversely, the workload and cost of monitoring, assessing and reviewing data for increasingly lengthy lists of substances is not sustainable, nor realistic. Screening is one way of identifying new candidates for future inclusion in monitoring programmes. Box 2.6 describes the process of the Danish approach to selecting pharmaceuticals for screenings. Switzerland has prioritised five indicator substances to reduce analytical costs of monitoring for an extensive list of CECs (Box 2.7).

Box 2.6. Monitoring and screening programme for pharmaceuticals, Denmark

In Demark, a national monitoring programme (NOVANA) governed by the Danish Environmental Protection Agency monitors EU priority substances and national river basin-specific substances in water. In addition to the NOVANA programme, Denmark also performs annual screening projects with the aim of identifying new candidates for future inclusion in the national monitoring programme. This approach is considered cost effective in order to collect knowledge of the presence of specific substances in the aquatic environment. Two pharmaceutical screenings have been performed, in 2008 and 2015. The following assumptions and requirements were used in order to select pharmaceuticals for the screenings:

- Pharmaceuticals specifically mentioned in the scientific literature as a potential high environmental risk
- Pharmaceuticals representing various applications
- Pharmaceuticals in high use in primary (hospital) and secondary (outpatient) care
- Pharmaceuticals that could be reliably analysed with available monitoring technologies.

The 2015 study selected and screened for 27 human pharmaceuticals that had not previously been included in the national monitoring programme (see Table below). Samples were taken from WWTPs (inlet, outlet and sludge) and freshwater and marine water bodies receiving wastewater discharges. A total of 66 samples were analysed within the project.

Table 2.2. Pharmaceuticals selected for screening in 2015, Denmark Pharmaceuticals selected for screening in 2015, Denmark

Pharmaceuticals in the screening	Application (therapeutic use)
Capecitabine, Tamoxifen	Anti-cancer
Carbamazepine, Citalopram, Codeine, Fen-tanyl, Propofol, Tramadol	Central nervous system agents
Miconazol	Antifungal
Amiloride, Atenolol, Bisoprolol, Losartan, Metoprolol, Propranolol	Cardiovascular agents
Amoxicillin, Azithromycin, Ciprofloxacin, Clarithromycin, Erythromycin, Fluconazol, Roxythromycin	Antibiotics
Diclofenac, Ibuprofen, Ketoprofen, Naproxen	NSAIDs
Norethisterone	Synthetic progestin

Based on the screening results, the following pharmaceuticals were included in the NOVANA programme: Azithromycin, Clarithromycin, Carbamazepin, Citalopram, Ibuprofen, Naproxen, Tramadol, Propranolol and Diclofenac. No legislation changes or determination of EQN for these substances has yet been carried out. The need for such actions will be further evaluated when the programme generates

---

more data for these pharmaceuticals. The 2015 screening, funded by the Danish Environment Agency, cost € 95 000. This case study highlights the importance of having knowledge from drug databases about which pharmaceutical substances are being used, where they are used (outpatient or hospital), and in what quantities in order to rank and select pharmaceuticals for screenings.

Source: Summary of case study provided by Henrik Søren Larsen, Ministry of Environment and Food, Environmental Protection Agency, Denmark

---

Box 2.7. Identifying and prioritising indicator substances for CECs monitoring, Switzerland

Switzerland has prioritised five indicator substances to reduce analytical costs of monitoring for an extensive list of CECs. Out of a total of 250 substances (pharmaceuticals, pesticide and transformation products) identified in Swiss rivers, 47 indicator substances were identified through a selection process based on five criteria: i) partitioning of substances between water and solid phase; ii) persistence in the aquatic environment; iii) toxicity; iv) concentration patterns (continuous, periodic or intermittent); and v) probability of detecting a substance in surface waters.

To reduce the analytical costs for monitoring all 47 compounds, a subgroup of five indicator compounds was identified to be included in sampling programmes: carbamazepine (anticonvulsant or anti-epileptic drug), diclofenac (nonsteroidal anti-inflammatory drug), sulfamethoxazole (antibiotic), mecoprop (herbicide) and benzotriazole (anticorrosive agent). All of these substances can be measured with the same analytical method and are detectable in more than 90 % of all domestic WWTP effluents in Switzerland.

Source: (Götz, Kase and Hollender, 2011[39])

In order to overcome the almost infinite number of combination of pharmaceuticals (and their metabolites and transformation products) in the environment, novel methods and alternative holistic approaches are being developed. These emerging monitoring technologies may be overtaking the capacity of governments to react and put adequate responses in place. Some countries are revising monitoring programmes to include the effects of mixtures. The following section takes stock of recent advances in monitoring (including integrated approaches) and modelling that can provide an alternative to assessing APIs in the environment individually.

2.4. Advances in water quality monitoring and potential benefits for risk assessments and water quality policy making

The sheer number of potentially harmful pharmaceuticals and other emerging pollutants challenges traditional chemical monitoring efforts, and consequently there is a good chance that adverse impacts from unknown or unexpected chemicals and mixtures on aquatic communities and human health remain unrecognised. The problem is aggravated by analytical detection limits that may be too high for detecting chemicals at or below their predicted no-effect-concentrations (PNEC). Finally, a better understanding is required of how to link early biological responses to chemical exposure detectable in bioassays and biomarkers to ecological responses at the population and community level (Brack et al., 2015[40]).

A number of advances in water quality monitoring and modelling techniques can assist with some of these challenges, providing alternatives to the costly traditional spot (grab) sampling and chemical analysis approach. There is value in moving towards approaches which aim at identifying and anticipating new environmental pollutants as early as possible, i.e. before they become pervasive in the aquatic environment and before major health or economic consequences are felt (Daughton, 2004[38]). It is particularly important to discover impacts related to environmental contaminants before they occur on a population level, because damage at the population and ecosystem level can take a long time to repair.

New methods including non-target screening and suspected-target screening can be utilised to gain knowledge about the occurrence of so-called "known unknowns" and "unknown unknowns" of pollutants in water bodies[2]. Box 2.8 illustrates an example of non-target and suspected-target screening in Korea.

Box 2.8. Prioritisation of pharmaceuticals via suspect and non-target screening, Korea

The Yeongsan River is one of four major river basins in Korea. It is the most water scarce basin and has suffered from declining water quality from an increase in diffuse urban and agricultural pollution and toxic point source discharges (OECD, 2018[41]).

In a study by Park et al. (2018[42]), pharmaceuticals and personal care products (PPCPs) in the Yeongsan River, Korea were prioritised using suspect and non-target analysis by Liquid chromatography–high resolution mass spectrometry (LC-HSMS) (QExactive plus Orbitrap) followed by semi-quantitative analysis to confirm the priority of PPCPs.

The screening identified more than 50 PPCPs, of which 26 could be confirmed with reference standards. The confirmed substances were prioritised based on a scoring and ranking system. Twelve additional substances not included in the first ranking were semi-quantitatively analysed. In the final prioritisation list, carbamazepine, metformin and paraxanthine shared first-ranking place, followed by caffeine, cimetidine, lidocaine, naproxen, cetirizine, climbazole, fexofenadine, tramadol, and fluconazole. The authors suggest that these 12 PPCPs are the most highly exposable substances, and should be considered in future water monitoring of the Yeongsan River (Park et al., 2018[42]).

Sources: (Park et al., 2018[42]; OECD, 2018[41])

Effect-based monitoring can be a valuable approach to detect effects caused by contaminants at an earlier stage than a substance-by substance chemical monitoring approach (Wernersson et al., 2015[43]). They can also take into account the overall response from co-exposure to multiple, bioavailable pharmaceuticals and chemicals in the environment, including on different levels of biological organisation, such as community, population, individual and/or suborganism level (EC, 2014[44]). The European Chemical Monitoring and Emerging Pollutants sub-working group recommends the continuation of such monitoring in relation to: i) the detection and evaluation of effects caused by mixtures of pollutants, ii) as screening tools as part of ERAs to aid in the prioritisation of water bodies, iii) to establish early-warning systems, and iv) to provide additional support in water and sediment quality assessment, complementary to conventional chemical monitoring under the WFD (EC, 2014[44]) (Wernersson et al., 2015[43]). For more information on the state of the art of aquatic effect-based monitoring tools, refer to (EC, 2014[44]).

An overview of the various monitoring approaches, differentiated by chemical monitoring and effect-based monitoring, and their advantages and disadvantages are presented in Table 2.3.

Table 2.3. An overview of the different monitoring approaches for pharmaceuticals in water, and their advantages and disadvantages

Monitoring approach	Description	Advantages	Disadvantages	Indicative, relative cost
Chemical monitoring				↗
Spot-sampling and target chemical Analysis	A discrete sample taken at one point in time and location, followed by laboratory analysis for a target chemical.	Appropriate for pharmaceuticals with known hazard.	Costly and labour-intensive. Episodic pollution events can be missed Fails to account for the bioavailability.	
Biosensors	An analytical device, used for the detection of a chemical substance that combines a biological component with a physicochemical detector.	Enables real-time *in situ* monitoring. Rapid, simple, low-cost techniques. Low number of samples required for quantification Valuable as an early warning signal.	Difficult to assess the impacts of exposure and the mixture effects of multiple contaminants on the ecosystem. Requires prior knowledge about the type of substance to be monitored for. Low selectivity (in most cases, exception being antibody-based sensors), low detection limits, risk of contamination with other microorganisms. Often requires skilled operators at the research stage, or more expensive user-friendly commercialised biosensors.	↘
Non-target and suspected screenings	Advanced techniques that often employ high resolution mass spectrometry and liquid chromatography to either match unknown sample features to compounds within spectral and/or spectra-less databases (suspect screening), or elucidate structures of unknowns that may not be contained in a database (non-target screening).	Measures a large amount of chemical features Useful to identify chemical structures of transformation products Does not require priori information on the compounds to be detected.	Non-target identification is a very time-intensive process. Unidentified substances remain "unknowns". Requires highly-trained personnel and expensive equipment. Can be difficult to distinguish between those signals worthwhile following up, and those that are just 'noise'.	↗
Passive sampling	Technique involving the use of a collecting medium, such as a man-made device or biological organism, to accumulate chemical pollutants in the environment over time. Also known as diffuse sampling.	Can effectively concentrate pollutants compared to spot sampling. Gives information about source, sink and patterns of pollution at different locations. Provides time-weighted-average and equilibrium concentrations over the deployment time, rather than a snap shot at one moment. Accounts for bioavailable fractions. Low cost, non-mechanical, easy to deploy, little/no maintenance. Low number of samples required for quantification.	Does not identify unknowns. Requires careful pre-calibration. No standardised guidance. Performance can be affected by environmental variables, such as flow rate, temperature, pH, salinity and biofouling.	↘
Effect-based (ecological) monitoring				↗
Ecological indicators	Biological assemblages or taxa that by their presence, condition or numbers indicate something about the state of the environment. Salmon and mayflies are well-known ecological indicators of the health of rivers.	Assesses the overall response from co-exposure to multiple, bioavailable chemicals (including transformation products), at different biological levels. Provides a highly integrated and relevant response.	Difficult to identify underlying causes and key events of the adverse effect of the ecosystem. Difficult to find reference conditions. Can be time-consuming and difficult to implement in a cost-effective manner without prior knowledge of what pharmaceuticals and effects (and in what biota) should be looked for.	↗
Bioassays (*in vitro*)	Test that measures the joint biological effect of all active chemicals in an environmental sample under defined laboratory conditions at the subcellular level, such as receptor	Monitors the overall biological activity and the potential mechanism of toxicity. Highly sensitive. Low cost, appropriate as rapid screening tool, can be used to identify and track	Limited to the targeted biological system and endpoints. No harmonised effect-based trigger values. Difficult to interpret results when they do	↘

Monitoring approach	Description	Advantages	Disadvantages	Indicative, relative cost
	activation and DNA damage. The OECD Test Guidelines Programme identifies new *in vitro* test methods that are candidates to become part of OECD Test Guidelines (OECD, 2018[45]).	pollution sources and water bodies that need require further investigations. Provides benchmarking of mixture effects Can be used to identify toxic fractions and provide guidance for the identification of causative agents.	not necessarily imply an adverse effect in exposed whole organisms (i.e. difficult to translate *in vitro* responses to *in vivo* effects).	
Bioassays (*in vivo*)	Test that measures the joint biological effect of all active chemicals in an environmental sample under defined laboratory conditions at the individual level. An example is the fish embryo acute toxicity test, adopted by the OECD as Test Guideline N.236, which is based on individual exposure of eggs to evaluate the embryotoxicity of samples with the aim to detect contaminants, including pharmaceuticals. Sometimes tests are performed in the field (*in situ* bioassays).	Provides broad spectrum analysis of the effects of a variety of substances (both known and unknown) and different types of toxicity to whole living organisms.	Costly and time-intensive. Often limited to the targeted biological endpoints.	↗
Biomarkers	Measure of biological alterations (e.g. biochemical, physiological, histological, or morphological changes) at cellular or individual levels, measured in organisms sampled in the field in a specific location (e.g. by collecting blood samples from mussels or fish). There are biomarkers of exposure (e.g. measured concentration of a target chemical in blood or the liver), which inform about the quality and/or quantity of exposure, and biomarkers of effects, which allow statements about effects and the health status of exposed organisms (e.g. genotoxicity, effects on the immune system or reproduction). The OECD Test Guidelines Programme identifies new biomarker endpoints (OECD, 2018[45]).	Measures biological alterations at cellular or individual levels in organisms at specific field locations. Identifies the impact from substances or mixtures of substances not previously identified to be of concern, and identify regions of decreased environmental quality. Provides benchmarking of mixture effects. Valuable as an early warning signal.	Difficult to establish linkage between alterations and adverse effects at community or population level. Interpreting the data can be challenging. Very few biomarkers can be linked directly to exposures to specific classes of chemicals.	↗
Omics	High-throughput molecular profiling technologies, such as genomics, metagenomics, transcriptomics, proteomics, metabolomics and metabonomics. Used to explore the roles, relationships, and actions of the various types of molecules that make up the cells of an organism.	Provides unique opportunities for the discovery of a new generation of biomarkers of exposure and disease risk. Can identify unexpected effects, monitor biological effects at the genome scale. Can be used to develop molecular biomarkers of exposure as early signals to predict effects. Provides information about the mechanism of toxicity and a basis for extrapolation of the effects across species, and effects on the whole ecosystem. Provides benchmarking of mixture effects.	Early stage of development. Lack of standardisation and guidance on best practices. Difficulties in interpreting data. Requires highly-trained personnel and expensive equipment.	↗

Notes: '↗' denotes relatively costly; '↘' denotes relatively inexpensive.

Sources: (Ejeian et al., 2018[46]) (Hernandez-Vargas et al., 2018[47]) (Chouler et al., 2015[48]) (Letzel, 2014[49]) (Novák et al., 2018[50]) (Brack et al., 2017[51]) (Anderson et al., 2012[52]) (EC, 2014[44]) (Wernersson et al., 2015[43]) (Escher et al., 2018[53]) (van der Oost et al., 2017[54]) (Könemann et al., 2018[55]) (Kase et al., 2018[56]) (Ekman et al., 2013[29]) (Leung, 2018[57]).

2.4.1. The value of integrating monitoring approaches

No single method or combination of methods is able to meet all divergent monitoring purposes (Altenburger et al., 2019[58]). However, employing various monitoring approaches (Table 2.3) together, utilising their various strengths, can provide a more holistic understanding of pharmaceuticals in the environment and their environmental effects (Figure 2.2) (Ekman et al., 2013[29]; Petrie, Barden and Kasprzyk-Hordern, 2015[59]) (Altenburger et al., 2019[58]).

The use of effect-based tools, passive sampling for bioaccumulative chemicals and an integrated strategy for prioritisation of contaminants, accounting for knowledge gaps, is advocated to improve and advance monitoring (Ekman et al., 2018[60]) (Brack et al., 2015[40]). Biomarkers can be effectively incorporated with other diagnostic markers of fish health, and also with analytical chemical monitoring approaches to provide evidence for the contributions of chemical exposures (Hook, Gallagher and Batley, 2014[61]). Integrating Adverse Outcome Pathways and omics would be a valuable practical tool to discover low-dose effects of substances, either individually or as a mixture (Leung, 2018[57]). In the OECD Conceptual Framework for Testing and Assessment of Endocrine Disruptors, *in vitro* assays are recommended to provide data on selected endocrine mechanisms and pathways before the substance eventually undergoes further investigations *in vivo* (OECD, 2018[62]; OECD, 2018[63]).

Figure 2.2. The components in integrated monitoring approaches and the key strengths and limitations of each approach

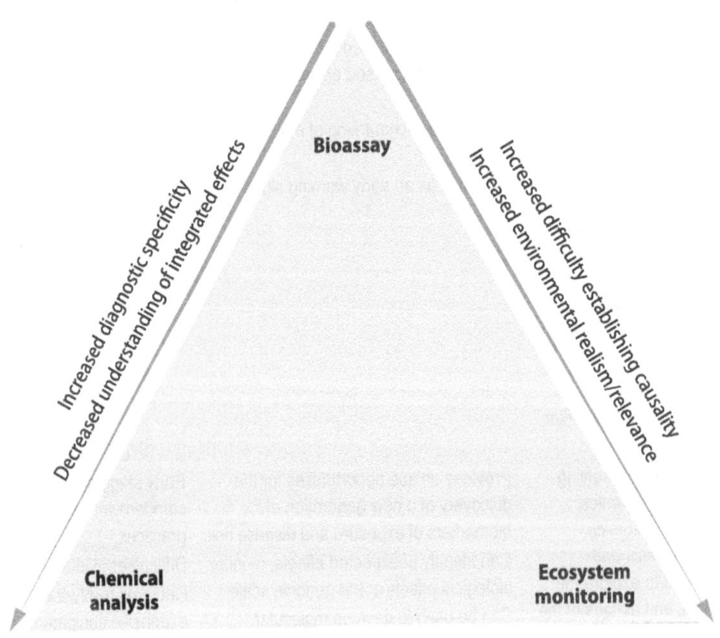

Source: (Ekman et al., 2013[29])

There are several integrated monitoring programmes in OECD countries. For example, as part of the U.S. Great Lakes Restoration Initiative, an integrated approach including chemical analysis, bioassays (*in vitro, in vivo* or *in situ*) and ecosystem monitoring is used to monitor pharmaceuticals and other emerging pollutants, which are identified as one of the highest priority stressors in the lakes (Ekman et al., 2013[29]). Also in the US, effect-based testing, combined with analytical chemistry data, is used to determine discharge permitting requirements (Box 2.9). In the Netherlands, passive sampling is used in combination with *in situ, in vivo* and *in vitro* bioassays to assess the impacts of wastewater discharge on water quality using the SIMONI (smart integrated monitoring) approach (van der Oost et al., 2017[64]). A Swiss study reveals the benefits of using chemical analysis, bioassays and mixture toxicity modelling to determine the impact of mixture effects and multiple sources (both point and diffuse) on instream water quality (Box 2.10).

Box 2.9. Whole Effluent Toxicity testing, US: A combination of testing methods to evaluate toxicity of wastewater

In the US, Whole Effluent Toxicity (WET) testing is used to determine the aggregate acute and short-term chronic toxicity of wastewater effluent on aquatic organisms. It is one way the EPA Clean Water Act's prohibition of the discharge of toxic pollutants in toxic amounts is implemented. WET tests measure the effects of wastewater on specific test organisms' ability to survive, grow and reproduce. The WET test methods consist of exposing living aquatic organisms (plants, vertebrates and invertebrates) to various concentrations of a sample of wastewater, usually from a facility's effluent stream. WET tests are used by the National Pollutant Discharge Elimination System permitting authority to determine whether a facility's permit will need to include WET requirements.

Two WET test manuals describe test procedures for effluents and receiving waters and include guidelines on test species selection and mobile toxicity test laboratory design. When toxicity is measured, the next step is to use the Toxicity Identification Evaluation (TIE) process to identify and reduce the toxicity. The TIE is a three-phase process that characterises, identifies, and confirms the cause or causes of toxicity. Once the identification/isolation process has confirmed the potential cause of toxicity, the next step is to determine what needs to be done to reduce or treat the chemical or chemicals causing toxicity in the effluent. *In vitro* assays are being used to screen water samples for the presence of compounds that activate specific cellular pathways (e.g. activating an estrogen-dependent pathway) or that target a particular cell type (e.g. bladder). Omics endpoints, measured in exposed organisms, represent the immediate and early response that underlie the development of adverse outcomes. Signatures from these new assays can be combined with traditional apical endpoints, and targeted and non-targeted analytical chemistry data, to strengthen the weight of evidence of biologically meaningful exposure and to identify causative agents.

Source: (US EPA, 2017[65])

Box 2.10. Integrated monitoring to evaluate the contribution of wastewater effluent on CECs burden in small streams, Switzerland

In a study by Neale et al. (2017[66]), chemical analysis and bioassays were combined to assess the CEC burden in small streams upstream and downstream of three WWTPs in the Swiss Plateau. The chemical analysis included 405 chemicals and *in vitro* bioassays assessed the following endpoints: activation of the aryl hydrocarbon receptor, androgen receptor and oestrogen receptor; photosystem II inhibition; acetylcholinesterase inhibition; adaptive stress responses for oxidative stress, genotoxicity and inflammation; and estrogenic activity and developmental toxicity in zebrafish embryos. Mixture toxicity modelling was applied to assess the contribution of detected chemicals to the observed effect.

The results from this study highlight the importance of combining bioassays with chemical analysis to provide the whole picture of the CEC burden to water bodies. The *in vitro* assay provided information about the mixture effects, while the chemical analysis showed differences in the chemical pollution profiles along different sampling stations. For most bioassays, very little of the observed effects could be explained by the detected chemicals. While higher concentrations and effects were observed in samples downstream of effluent discharge compared to upstream samples, both chemical analysis and bioanalysis showed that diffuse sources upstream during low flow conditions significantly affect the water quality and aquatic ecosystem. Consequently, upgrading WWTPs alone will not completely reduce the concentration and effects of CECs in the Swiss Plateau.

Source: (Neale et al., 2017[66]).

Passive sampling, together with effect-based approaches, are being considered as potentially suitable tools that could be employed for monitoring of European water bodies in the implementation strategy of the EU Water Framework Directive (European Commission, 2015). While both approaches are often employed independently, their use in combination has been demonstrated previously in studies focusing on WWTP effluents and affected rivers (Creusot et al., 2013; Jalova et al., 2013; Jarosova et al., 2012).

While the ideal environmental monitoring system would routinely include all three components (chemical analysis, bioassays, ecosystem/effect-based monitoring), resource limitations and other practical constraints generally dictate where they are employed on a case-by case basis (Ekman et al., 2013[29]). For example, some sites have limited prior knowledge and information about the pollution burden and possible effects, other sites have low ecological health status, and others have known effects and exposure where the major sources needs to be identified. Ekman et al. (2013[29]) suggest a stepwise process to design and implement a strategic and integrated monitoring approach. The first step is to start with a problem formulation considering the existing information about the site, management goals and particular regulatory motivators. Once these basics are understood, strategic decisions about which specific monitoring tools should be employed are possible, which then allows for informed decision-making on what actions may (or may not) be required.

One of the challenges using alternative tools and data types is the interpretation of endpoints in the context of biological effects. Biological activity (e.g. measured *in vitro* or in biomarkers) does not necessarily constitute a hazard. The Adverse Outcome Pathways (AOP) concept (OECD, 2018[45]) was developed to address this issue and to make it possible to relate endpoints from *in vitro* bioassays and biomarkers to endpoints useful for risk assessment (e.g. of survival, reproduction, development, and growth) (Ankley et al., 2010[67]) (Ankley and Edwards, 2018[68]). It is a conceptual chain that links the exposure of contaminants to their cellular concentrations and molecular initiating events, via pathway disturbance and key events, to response at the cellular, organism and population or community levels. The framework can be used for different purposes, including to predict the effects of mixtures (Carusi et al., 2018[69]). AOPs increase the efficiency of chemical safety assessments, reduce the need for animal testing, and has received significant attention and use in the regulatory toxicology community (Carusi et al., 2018[69]). Recently, the AOP has been moving from a linear concept to a pathway network considering the idea of multiple causes for adverse effects (Escher et al., 2017[70]) (Knapen et al., 2015[71]).

Another alternative method to predict the toxicity of pharmaceuticals, is to group APIs that are structurally similar and may therefore cause similar adverse environmental effects. The OECD has developed the Integrated Approaches to Testing and Assessment (IATA, which can include AOPs), providing a framework and tools for data gathering to maximise the amount of information about risks of chemicals. For example, the OECD Quantitative Structure-Activity Relationships (QSAR) Toolbox (OECD, 2018[72]) is a software application intended to be used by governments, chemical industry and other stakeholders to fill gaps in (eco)toxicity data needed for assessing the hazards of chemicals. The Toolbox incorporates information and tools from various sources into a logical workflow. It enables the identification of new methods/profilers for grouping chemicals through *in vitro* bioassays.

2.5. The added value of system modelling

Because of the lack of systematic monitoring programmes for pharmaceuticals, modelling serves as a valuable and cost-effective basis for prioritisation, risk assessments, and to address data and knowledge gaps. Modelling the source-to-effect chain can be an effective tool to identify and target sources of pollution for monitoring. They can also be useful for predicting the exposure and impacts of pollution sources, pharmaceutical types and mixture toxicity, and the effectiveness of technical and policy solutions. Modelling can be a constructive starting point to understand and discuss the source and effects of pharmaceuticals with stakeholders, from which cost-effective solutions can be developed in cooperation. Ultimately, it would be of value to move towards developing modelling approaches of real world exposure where organisms continually face mixtures of multitudes of stressors, which vary over time in composition and concentrations (Daughton, 2004[38]).

Computational tools, like QSAR (Quantitative Structure Activity Relationship) models, can be used to screen large sets of chemicals in a short time, with the aim of ranking, highlighting and prioritising the most environmentally hazardous for focusing further experimental studies (Sangion and Gramatica, 2016[73]). The QSAR approach has been used to model the toxicity of pharmaceuticals, both regarding mixtures and in the assessment of unknown substances such as transformation products (Escher et al., 2006[74]) (Rastogi, Leder and Kümmerer, 2014[75]) (Mahmoud et al., 2014[76]). The OECD has recognised the potential for QSARs to reduce the costs of testing, reduce the need for animal testing and to strengthen chemical regulation (Fjodorova et al., 2008[77]).

Modelling sources of pharmaceutical pollution can be used in order to investigate to what degree individual sources (such as WWTPs) impact water quality. For example, in the Netherlands, nation-wide consumption-based hydrological modelling has given spatial insight on the impact of WWTP discharge on concentrations of pharmaceuticals in surface water bodies (Box 2.11). In Sweden, modelling has determined the exposure potential of some of the nation's top-used pharmaceuticals to the Baltic Sea (Box 2.12). Oldenkamp et al. (2018[78]) have developed a spatially explicit model (ePiE) which calculates concentrations of APIs in surface waters across Europe at a resolution of approximately 1 km. Such models can be helpful in prioritising APIs and spatially assessing the risks.

> ### Box 2.11. Defining impact of wastewater treatment plants on susceptible functions, Netherlands
>
> A nation-wide modelling and ranking exercise was undertaken in the Netherlands to investigate and prioritise which of their 345 WWTPs should be upgraded to reduce the impact of pharmaceuticals on receiving water bodies (in particular to EU nature protection areas) and the risk to raw drinking water sources. The model was based on two components: i) a water quality model representing the Dutch surface water network and its key hydrological features; and ii) a consumption-based emission model to project the loads from WWTPs to receiving rivers during both low and high discharge conditions. Two pharmaceuticals with different characteristics (carbamazepine and ibuprofen) underwent a detailed spatial analysis.
>
> The vast majority of the total impact of all Dutch WWTPs, during both high and low discharge conditions, was attributed to 19% of the WWTPs with regard to the drinking water function, and to 39% of the WWTPs with regard to the nature protected areas function. The model thus provides a spatially smart and cost-effective way to identify and prioritise WWTP upgrades to improve water quality and reduce adverse environment effects.
>
> *Source:* (Coppens et al., 2015[79]).

> ## Box 2.12. Modelling to predict occurrence of multiple pharmaceuticals in Swedish surface waters and their release to the Baltic Sea
>
> A Swedish study provides assessments of the emissions and occurrence of 54 pharmaceuticals in surface water and their releases to the Baltic Sea. Lindim et al. (2017[80]) used the STREAM-EU model to predict the exposure, fate and long-range transport of 54 human pharmaceuticals (selected from the 200 most consumed pharmaceuticals in Sweden in 2011) in all Swedish basins draining to the Baltic Sea and Danish Strait. The model considered point sources (urban WWTPs) and diffuse sources (through WWTP sludge application to soils) as human pharmaceutical entry-pathways to surface water. The model was parametrised with the knowledge that 25.4% of the sludge generated from WWTPs in Sweden is used on agriculture soils, according to national statistics. The model was constructed by using data on emissions, hydrology, pH, air and water temperatures, and physico-chemical properties and computationally-estimated partitioning and degradation properties of the substances.
>
> The results showed a total flushing flow rate to the Baltic Sea of 42 ton/yr for the 54 pharmaceuticals combined. Of the 54 studied pharmaceuticals, 35 (65%) had predicted annual flushes to the sea higher that 1 kg/yr; the highest of which were for metformin (27 ton/yr), paracetamol (6.9 ton/yr) and ibuprofen (2.33 ton/yr). In the Stockholm urban area, 25 drugs had predicted concentrations higher than 1 ng/L, with 17 above 10 ng/L. Modelled results were in good agreement with monitored/measured values, with agreements of $r^2 = 0.62$ and $r^2 = 0.95$ for concentrations and for disposed amounts to sea, respectively. Persistence and hydrophobicity of the pharmaceuticals determined the long-range transport and exposure potential. The model indicated that piperacillin, lorazepam, metformin, hydroxycarbamide, hydrochlorothiazide, furosemide and cetirizine have high potential to reach the sea.
>
> *Source*: (Lindim et al., 2016[81]) (Lindim et al., 2017[80])

Modelling of pharmaceuticals at the global scale has also been attempted. Oldenkamp et al. (2019[82]) modelled the aquatic risks (expressed as the PEC/PNEC ratios) of antibiotics carbamazepine and ciprofloxacin in 449 aquatic ecoregions worldwide over the period 2005 to 2015. The study combined spatially explicit chemical fate and effect modelling with predictions of pharmaceutical consumption. The results showed exposure of antibiotics to freshwater ecosystems increased 10-20 fold over the last 20 years. Aquatic risks due to carbamazepine exposure were typically low, although more densely populated (e.g. western and central Europe) and/or arid ecoregions (e.g. Arabian and Californian peninsulas) showed higher risk. Risks for ciprofloxacin were found to be much higher and more widespread (Oldenkamp, Beusen and Huijbregts, 2019[82]).

Despite the advantages of modelling, it is pertinent to note that models only provide an estimation of reality and therefore there are always uncertainties. Source and environmental fate models require detailed information regarding pharmaceutical inputs on emission rates, partitioning properties, as well as degradation rates to accurately predict environmental partitioning and transport (Lindim et al., 2017[80]). Measured partitioning and degradation rates are still scarce for the majority of pharmaceuticals. Furthermore, many pharmaceuticals are polar and mobile, meaning that they are difficult to model as they behave differently from, for example, PBT-substances (Reemtsma et al., 2016[83]).

Finally, it is important to note that improving knowledge is not a pollution reduction measure in itself. Governments should take advantage of alternative innovative monitoring and modelling technologies to minimise costs, and prioritise substances and water bodies of highest concern, but they should not wait for exact science before taking action. Since pharmaceuticals are ubiquitous, the goal is to strive towards a non-toxic environment - not a non-chemical environment. Chapter 3 looks at opportunities to design and implement innovative and effective policies to reduce pharmaceuticals in the environment.

References

Ågerstrand, M. et al. (2015), "Improving environmental risk assessment of human pharmaceuticals", *Environmental Science and Technology*, Vol. 49/9, pp. 5336-5345, http://dx.doi.org/10.1021/acs.est.5b00302. [14]

Ågerstrand, M. et al. (2017), "An academic researcher's guide to increased impact on regulatory assessment of chemicals", *Environmental Science: Processes and Impacts*, Vol. 19/5, pp. 644-655, http://dx.doi.org/10.1039/c7em00075h. [26]

Altenburger, R. et al. (2019), "Future water quality monitoring: improving the balance between exposure and toxicity assessments of real-world pollutant mixtures", *Environmental Sciences Europe*, Vol. 31/1, p. 12, http://dx.doi.org/10.1186/s12302-019-0193-1. [58]

Anderson, P. et al. (2012), *Monitoring Strategies for Chemicals of Emerging Concern (CECs) in California's Aquatic Ecosystems Recommendations of a Science Advisory Panel.* [52]

Ankley, G. et al. (2010), "Adverse outcome pathways: A conceptual framework to support ecotoxicology research and risk assessment", *Environmental Toxicology and Chemistry*, Vol. 29/3, pp. 730-741, http://dx.doi.org/10.1002/etc.34. [67]

Ankley, G. and S. Edwards (2018), "The adverse outcome pathway: A multifaceted framework supporting 21st century toxicology", *Current Opinion in Toxicology*, Vol. 9, pp. 1-7, http://dx.doi.org/10.1016/J.COTOX.2018.03.004. [68]

aus der Beek, T. et al. (2016), "Pharmaceuticals in the environment-Global occurrences and perspectives", *Environmental Toxicology and Chemistry*, Vol. 35/4, pp. 823-835, http://dx.doi.org/10.1002/etc.3339. [30]

Australian Government (2008), *Australian guidelines for water recycling: managing health and environmental risks (phase 2). Augmentation of drinking water supplies.*, National Health and Medical Research Council, Canberra. [37]

Backhaus, T. (2014), "Medicines, shaken and stirred: A critical review on the ecotoxicology of pharmaceutical mixtures", *Philosophical Transactions of the Royal Society B: Biological Sciences*, Vol. 369/1656, http://dx.doi.org/10.1098/rstb.2013.0585. [23]

BIO Intelligence Service (2013), *Study on the environmental risks of medicinal products.* [24]

Brack, W. et al. (2015), "The SOLUTIONS project: Challenges and responses for present and future emerging pollutants in land and water resources management", *Science of the Total Environment*, Vol. 503-504, pp. 22-31, http://dx.doi.org/10.1016/j.scitotenv.2014.05.143. [40]

Brack, W. et al. (2017), "Towards the review of the European Union Water Framework management of chemical contamination in European surface water resources", *Science of the Total Environment*, Vol. 576, pp. 720-737, http://dx.doi.org/10.1016/j.scitotenv.2016.10.104. [51]

Caneva, L. et al. (2014), "Critical review on the Environmental Risk Assessment of medicinal products for human use in the centralised procedure", *Regulatory Toxicology and Pharmacology*, Vol. 68/3, pp. 312-316, http://dx.doi.org/10.1016/J.YRTPH.2014.01.002. [3]

Carusi, A. et al. (2018), "Harvesting the promise of AOPs: An assessment and recommendations", *Science of the Total Environment*, Vol. 628-629, pp. 1542-1556, http://dx.doi.org/10.1016/j.scitotenv.2018.02.015. [69]

Chapman, J. et al. (2017), "Three methods for integration of environmental risk into the benefit-risk assessment of veterinary medicinal products", *Science of the Total Environment*, Vol. 605-606, pp. 692-701, http://dx.doi.org/10.1016/j.scitotenv.2017.06.205. [10]

Chouler, J. et al. (2015), "Water Quality Monitoring in Developing Countries; Can Microbial Fuel Cells be the Answer?", *Biosensors*, Vol. 5/3, pp. 450-470, http://dx.doi.org/10.3390/bios5030450. [48]

Cleuvers, M. (2008), "Chronic Mixture Toxicity of Pharmaceuticals to Daphnia – The Example of Nonsteroidal Anti-Inflammatory Drugs", in *Pharmaceuticals in the Environment*, Springer Berlin Heidelberg, Berlin, Heidelberg, http://dx.doi.org/10.1007/978-3-540-74664-5_17. [20]

Coppens, L. et al. (2015), "Towards spatially smart abatement of human pharmaceuticals in surface waters: Defining impact of sewage treatment plants on susceptible functions", *Water Research*, Vol. 81, pp. 356-365, http://dx.doi.org/10.1016/j.watres.2015.05.061. [79]

Daughton, C. (2004), "Non-regulated water contaminants: Emerging research", *Environmental Impact Assessment Review*, Vol. 24/7-8, pp. 711-732, http://dx.doi.org/10.1016/j.eiar.2004.06.003. [38]

EC (2018), *Updated surface water Watch List adopted by the Commission*, https://ec.europa.eu/jrc/en/science-update/updated-surface-water-watch-list-adopted-commission (accessed on 25 January 2019). [32]

EC (2014), *Technical report on aquatic effect-based monitoring tools*, European Commission, http://dx.doi.org/10.2779/7260. [44]

Ejeian, F. et al. (2018), "Biosensors for wastewater monitoring: A review", *Biosensors and Bioelectronics*, Vol. 118, pp. 66-79, http://dx.doi.org/10.1016/J.BIOS.2018.07.019. [46]

Ekman, D. et al. (2013), "Biological Effects–Based Tools for Monitoring Impacted Surface Waters in the Great Lakes: A Multiagency Program in Support of the Great Lakes Restoration Initiative", *Environmental Practice*, Vol. 15/4, pp. 409-426, http://dx.doi.org/10.1017/S1466046613000458. [29]

Ekman, D. et al. (2018), "Evaluation of targeted and untargeted effects-based monitoring tools to assess impacts of contaminants of emerging concern on fish in the South Platte River, CO", *Environmental Pollution*, Vol. 239, pp. 706-713, http://dx.doi.org/10.1016/J.ENVPOL.2018.04.054. [60]

EMA (2017), *Reflection paper on the authorisation of veterinary medicinal products containing (potential) persistent, bioaccumulative and toxic (PBT) or very persistent and very bioaccumulative (vPvB) substances*, European Medicines Agency , London, http://www.ema.europa.eu/docs/en_GB/document_library/Scientific_guideline/2017/05/WC500228196.pdf. [12]

EMA (2009), *Recommendation on the Evaluation of the Benefit-Risk Balance of Veterinary Medicinal Products. EMEA/CVMP/248499/2007*, European Medicines Agency Veterinary Medicines and Inspections, London, http://www.ema.europa.eu/docs/en_GB/document_library/Other/2009/10/WC500005264.pdf. [9]

EMA (2006), *Guideline on the environmental risk assessment of medicinal products for human use. Doc. Ref. EMEA/CHMP/SWP/4447/00 corr 2*, European Medicines Agency Pre-Authorisation Evaluation of Medicines for Human Use, London. [1]

EMA (2004), *Guideline on Environmental Impact Assessment for Veterinary Medicinal Products Phase II. CVMP/VICH/790/03¬FINAL*, European Medicines Agency Veterinary Medicines and Inspections, London,
http://www.ema.europa.eu/docs/en_GB/document_library/Scientific_guideline/2009/10/WC500004393.pdf. [8]

Escher, B. et al. (2018), "Effect-based trigger values for in vitro and in vivo bioassays performed on surface water extracts supporting the environmental quality standards (EQS) of the European Water Framework Directive", *Science of The Total Environment*, Vol. 628-629, pp. 748-765, http://dx.doi.org/10.1016/J.SCITOTENV.2018.01.340. [53]

Escher, B. et al. (2006), "Comparative ecotoxicological hazard assessment of beta-blockers and their human metabolites using a mode-of-action-based test battery and a QSAR approach", *Environmental Science and Technology*, Vol. 40/23, pp. 7402-7408, http://dx.doi.org/10.1021/es052572v. [74]

Escher, B. et al. (2017), "From the exposome to mechanistic understanding of chemical-induced adverse effects", *Environment International*, Vol. 99, pp. 97-106, http://dx.doi.org/10.1016/j.envint.2016.11.029. [70]

European Commission (2017), *Strategies against chemical pollution of surface waters*, http://ec.europa.eu/environment/water/water-dangersub/ (accessed on 11 September 2018). [34]

FDA (2016), *Environmental Assessment: Questions and Answers Regarding Drugs With Estrogenic, Androgenic, or Thyroid Activity Guidance for Industry*, U.S. Department of Health and Human Services Food and Drug Administration Center for Drug Evaluation and Research (CDER), Silver Spring, MD, http://www.fda.gov/Drugs/GuidanceComplianceRegulatoryInformation/Guidances/default.htm. [4]

FDA (1998), *Guidance for Industry Environmental Assessment of Human Drug and Biologics Applications*, Food and Drug Administration, https://www.fda.gov/downloads/Drugs/Guidances/ucm070561.pdf. [5]

Fjodorova, N. et al. (2008), "Directions in QSAR modeling for regulatory uses in OECD member countries, EU and in Russia", *Journal of Environmental Science and Health - Part C Environmental Carcinogenesis and Ecotoxicology Reviews*, Vol. 26/2, pp. 201-236, http://dx.doi.org/10.1080/10590500802135578. [77]

Godoy, A. and F. Kummrow (2017), "What do we know about the ecotoxicology of pharmaceutical and personal care product mixtures? A critical review", *Critical Reviews in Environmental Science and Technology*, Vol. 47/16, pp. 1453-1496, http://dx.doi.org/10.1080/10643389.2017.1370991. [22]

Götz, C., R. Kase and J. Hollender (2011), *Micropollutants - Assessment concept for organic trace substances from municipal sewage [In German]*, Swiss Federal Institute of Aquatic Science and Technology (Eawag), Dübendorf. [39]

Gunnarsson, L. et al. (2019), "Pharmacology beyond the patient – The environmental risks of human drugs", *Environment International*, Vol. 129, pp. 320-332, http://dx.doi.org/10.1016/J.ENVINT.2019.04.075. [2]

Hernandez-Vargas, G. et al. (2018), "Electrochemical biosensors: A solution to pollution detection with reference to environmental contaminants", *Biosensors*, Vol. 8/2, http://dx.doi.org/10.3390/bios8020029. [47]

Hook, S., E. Gallagher and G. Batley (2014), "The role of biomarkers in the assessment of aquatic ecosystem health.", *Integrated environmental assessment and management*, Vol. 10/3, pp. 327-41, http://dx.doi.org/10.1002/ieam.1530. [61]

Kase, R. et al. (2018), "Screening and risk management solutions for steroidal estrogens in surface and wastewater", *TrAC Trends in Analytical Chemistry*, Vol. 102, pp. 343-358, http://dx.doi.org/10.1016/J.TRAC.2018.02.013. [56]

Knapen, D. et al. (2015), "The potential of AOP networks for reproductive and developmental toxicity assay development", *Reproductive Toxicology*, Vol. 56, pp. 52-55, http://dx.doi.org/10.1016/j.reprotox.2015.04.003. [71]

Könemann, S. et al. (2018), "Effect-based and chemical analytical methods to monitor estrogens under the European Water Framework Directive", *TrAC - Trends in Analytical Chemistry*, Vol. 102, pp. 225-235, http://dx.doi.org/10.1016/j.trac.2018.02.008. [55]

Kools, S. et al. (2008), "A ranking of European veterinary medicines based on environmental risks", *Integrated Environmental Assessment and Management*, Vol. 4/4, pp. 399-408, http://dx.doi.org/10.1897/IEAM_2008-002.1. [15]

Kortenkamp Andreas, T. and M. Faust (2009), *State of the Art Report on Mixture Toxicity*, Directorate General for the Environment: Report to the EU Commission, http://ec.europa.eu/environment/chemicals/effects/pdf/report_mixture_toxicity.pdf. [21]

Kümmerer, K. et al. (2016), "Antineoplastic compounds in the environment—substances of special concern", *Environmental Science and Pollution Research*, Vol. 23/15, pp. 14791-14804, http://dx.doi.org/10.1007/s11356-014-3902-8. [16]

Küster, A. and N. Adler (2014), "Pharmaceuticals in the environment: Scientific evidence of risks and its regulation", *Philosophical Transactions of the Royal Society B: Biological Sciences*, Vol. 369/1656, http://dx.doi.org/10.1098/rstb.2013.0587. [13]

Le Page, G. et al. (2017), "Integrating human and environmental health in antibiotic risk assessment: A critical analysis of protection goals, species sensitivity and antimicrobial resistance", *Environment International*, Vol. 109, pp. 155-169, http://dx.doi.org/10.1016/J.ENVINT.2017.09.013. [17]

Le Page, G. et al. (2019), "Variability in cyanobacteria sensitivity to antibiotics and implications for environmental risk assessment", *Science of The Total Environment*, Vol. 695, p. 133804, http://dx.doi.org/10.1016/J.SCITOTENV.2019.133804. [18]

Letzel, T. (2014), *Non-target screening, suspected-target screening and target screening – of technologies and philosophies, databases and crafts*, http://www.int.laborundmore.com/archive/555305/Non-target-screening,-suspected-target-screening-and-target-screening-%E2%80%93-of-technologies-and-philosophies,-databases-and-crafts.html (accessed on 11 September 2018). [49]

Leung, K. (2018), "Joining the dots between omics and environmental management", *Integrated Environmental Assessment and Management*, Vol. 14/2, pp. 169-173, http://dx.doi.org/10.1002/ieam.2007. [57]

Lindim, C. et al. (2017), "Model-predicted occurrence of multiple pharmaceuticals in Swedish surface waters and their flushing to the Baltic Sea", *Environmental Pollution*, Vol. 223, pp. 595-604, http://dx.doi.org/10.1016/J.ENVPOL.2017.01.062. [80]

Lindim, C. et al. (2016), "Evaluation of human pharmaceutical emissions and concentrations in Swedish river basins", *Science of the Total Environment*, Vol. 572, pp. 508-519, http://dx.doi.org/10.1016/j.scitotenv.2016.08.074. [81]

Little, J., C. Cleven and S. Brown (2011), "Identification of "Known Unknowns" Utilizing Accurate Mass Data and Chemical Abstracts Service Databases", *Journal of The American Society for Mass Spectrometry*, Vol. 22/2, pp. 348-359, http://dx.doi.org/10.1007/s13361-010-0034-3. [84]

Mahmoud, W. et al. (2014), "Identification of phototransformation products of thalidomide and mixture toxicity assessment: An experimental and quantitative structural activity relationships (QSAR) approach", *Water Research*, Vol. 49/1, pp. 11-22, http://dx.doi.org/10.1016/j.watres.2013.11.014. [76]

Marsland, T. and S. Roy (2016), "Groundwater Watch List: Pharmaceuticals Pilot Study. Monitoring Data Collection and Initial Analysis. For European Commission – and CIS Groundwater Working Group (WGGW)", Amec Foster Wheeler Associate, Shrewsbury, http://www.vlakwa.be/fileadmin/user_upload/20160629-File3-1600204_Pharm_Pilot_Study_GRW.doc. [33]

MPA (2015), *Miljöindikatorer inom ramen för nationella läkemedelsstrategin (NLS). [Environmental indicators under the national drugs strategy (NLS)] Rapport från CBL-kansliet, Läkemedelsverket*, Swedish Medicinal Product Agency, Uppsala, https://lakemedelsverket.se/upload/nyheter/2015/miljoindikatorer-rapport-NLS_2015-09-07.pdf. [35]

Neale, P. et al. (2017), "Integrating chemical analysis and bioanalysis to evaluate the contribution of wastewater effluent on the micropollutant burden in small streams", *Science of the Total Environment*, Vol. 576, pp. 785-795, http://dx.doi.org/10.1016/j.scitotenv.2016.10.141. [66]

Novák, J. et al. (2018), "Effect-based monitoring of the Danube River using mobile passive sampling", *Science of The Total Environment*, Vol. 636, pp. 1608-1619, http://dx.doi.org/10.1016/J.SCITOTENV.2018.02.201. [50]

OECD (2019), *Saving Costs in Chemicals Management: How the OECD Ensures Benefits to Society*, OECD Publishing, Paris, https://dx.doi.org/10.1787/9789264311718-en. [25]

OECD (2018), *Adverse Outcome Pathways, Molecular Screening and Toxicogenomics*, http://www.oecd.org/chemicalsafety/testing/adverse-outcome-pathways-molecular-screening-and-toxicogenomics.htm (accessed on 27 January 2019). [45]

OECD (2018), *Considerations for Assessing the Risks of Combined Exposure to Multiple Chemicals, Series on Testing and Assessment No. 296, Environment, Health and Safety Division, Environment Directorate*. [27]

OECD (2018), *Managing the Water-Energy-Land-Food Nexus in Korea: Policies and Governance Options*, OECD Studies on Water, OECD Publishing, Paris, https://dx.doi.org/10.1787/9789264306523-en. [41]

OECD (2018), *OECD Work Related to Endocrine Disrupters*, http://www.oecd.org/env/ehs/testing/oecdworkrelatedtoendocrinedisrupters.htm (accessed on 27 January 2019). [63]

OECD (2018), *Revised Guidance Document 150 on Standardised Test Guidelines for Evaluating Chemicals for Endocrine Disruption*, OECD Series on Testing and Assessment, No. 150, OECD Publishing, Paris, https://dx.doi.org/10.1787/9789264304741-en. [62]

OECD (2018), *The OECD QSAR Toolbox*, http://www.oecd.org/chemicalsafety/risk-assessment/oecd-qsar-toolbox.htm (accessed on 27 January 2019). [72]

Oldenkamp, R., A. Beusen and M. Huijbregts (2019), "Aquatic risks from human pharmaceuticals—modelling temporal trends of carbamazepine and ciprofloxacin at the global scale", *Environmental Research Letters*, Vol. 14/3, p. 034003, http://dx.doi.org/10.1088/1748-9326/ab0071. [82]

Oldenkamp, R. et al. (2018), "A High-Resolution Spatial Model to Predict Exposure to Pharmaceuticals in European Surface Waters: ePiE", *Environmental science & technology*, Vol. 52/21, pp. 12494-12503, http://dx.doi.org/10.1021/acs.est.8b03862. [78]

Park, N. et al. (2018), "Prioritization of highly exposable pharmaceuticals via a suspect/non-target screening approach: A case study for Yeongsan River, Korea", *Science of the Total Environment*, Vol. 639, pp. 570-579, http://dx.doi.org/10.1016/j.scitotenv.2018.05.081. [42]

Petrie, B., R. Barden and B. Kasprzyk-Hordern (2015), "A review on emerging contaminants in wastewaters and the environment: Current knowledge, understudied areas and recommendations for future monitoring", *Water Research*, Vol. 72, pp. 3-27, http://dx.doi.org/10.1016/J.WATRES.2014.08.053. [59]

Rastogi, T., C. Leder and K. Kümmerer (2014), "Qualitative environmental risk assessment of photolytic transformation products of iodinated X-ray contrast agent diatrizoic acid", *Science of the Total Environment*, Vol. 482-483/1, pp. 378-388, http://dx.doi.org/10.1016/j.scitotenv.2014.02.139. [75]

Reemtsma, T. et al. (2016), "Mind the Gap: Persistent and Mobile Organic Compounds - Water Contaminants That Slip Through", *Environmental Science and Technology*, Vol. 50/19, pp. 10308-10315, http://dx.doi.org/10.1021/acs.est.6b03338. [83]

RIVM (2014), *Environmental risk limits for pharmaceuticals Derivation of WFD water quality standards for carbamazepine, metoprolol, metformin and amidotrizoic acid*, National Institute for Public Health and the Environment , Bilthoven. [36]

Sangion, A. and P. Gramatica (2016), "PBT assessment and prioritization of contaminants of emerging concern: Pharmaceuticals", *Environmental Research*, Vol. 147, pp. 297-306, http://dx.doi.org/10.1016/j.envres.2016.02.021. [73]

UBA (2018), *Recommendations for reducing micropollutants in waters. Background- April 2018*, German Environment Agency, Dessau-Roßlau. [11]

US EPA (2017), *Whole Effluent Toxicity Methods, Clean Water Act Analytical Methods*, United States Environmental Protection Agency, https://www.epa.gov/cwa-methods/whole-effluent-toxicity-methods (accessed on 25 June 2019). [65]

US EPA (2016), *Fourth Unregulated Contaminant Monitoring Rule, Monitoring unregulated drinking water contaminants*, https://www.epa.gov/dwucmr/fourth-unregulated-contaminant-monitoring-rule (accessed on 25 June 2019). [31]

van der Oost, R. et al. (2017), "SIMONI (Smart Integrated Monitoring) as a novel bioanalytical strategy for water quality assessment: Part II–field feasibility survey", *Environmental Toxicology and Chemistry*, Vol. 36/9, pp. 2400-2416, http://dx.doi.org/10.1002/etc.3837. [64]

van der Oost, R. et al. (2017), "SIMONI (smart integrated monitoring) as a novel bioanalytical strategy for water quality assessment: Part i–model design and effect-based trigger values", *Environmental Toxicology and Chemistry*, Vol. 36/9, pp. 2385-2399, http://dx.doi.org/10.1002/etc.3836. [54]

VICH (2018), *What is the role of VICH*, https://vichsec.org/what-is-the-role-of-vich.html (accessed on 11 September 2018). [6]

VICH (2000), *Environmental Impact Assessment (EIAs) for veterinary medicinal products (VMPs)- Phase I. VICH GL6*, The International Cooperation on Harmonisation of Technical Requirements for Registration of Veterinary Medicinal Products , Bruxelles. [7]

Wernersson, A. et al. (2015), "The European technical report on aquatic effect-based monitoring tools under the water framework directive", *Environmental Sciences Europe*, Vol. 27/1, pp. 1-11, http://dx.doi.org/10.1186/s12302-015-0039-4. [43]

WHO (2017), *Chemical mixtures in source water and drinking-water*, World Health Organisation, Geneva. [19]

WHO (2012), *Pharmaceuticals in Drinking Water*, World Health Organisation, Geneva. [28]

Notes

1 Communication from the Commission to the council: http://ec.europa.eu/transparency/regdoc/rep/1/2012/EN/1-2012-252-EN-F1-1.Pdf.

2 Unknowns" refer to a class of substances that cannot be categorised into known molecules or identified by standard evaluation methods (Little, Cleven and Brown, 2011[84]).

3 Emerging policy instruments for the control of pharmaceuticals in water

The chapter documents existing policy approaches to manage pharmaceutical residues in freshwater. The chapter draws inspiration from a range of OECD country case studies using various policy approaches - source-directed, use-orientated and end-of-pipe.

3.1. Key messages

While pharmaceuticals have undeniable benefits, countries struggle to address their potential impacts on drinking water quality, and ecosystem and human health. Pharmaceutical pollution challenges traditional water quality management, requiring new technologies in wastewater treatment and behavioural changes in industry, agriculture and health care sectors and society at large.

Current policy approaches to water quality protection are typically reactive; measures are adopted only when routine monitoring is in place and risks can be proven. Many country responses to the pharmaceutical problem have focussed on monitoring and end-of pipe measures (e.g. upgrading wastewater treatment plants and implementing public take-back programmes for unused and expired medicines). However, upstream, source-directed and use-orientated approaches are emerging, such as restrictions on the use of antibiotics in agriculture, green pharmacy and public environmental health campaigns.

3.2. Introduction

As the identification and understanding of pharmaceuticals in the environment improves, and attention from both the scientific community and public media increases, so does the need for policy intervention to reduce human and environmental risks. Various possibilities for the abatement of pharmaceuticals in the environment have received relatively little attention; many OECD countries and research institutions have focused on monitoring programmes, problem diagnosis, risk analysis and prioritisation of problematic pharmaceuticals.

This chapter looks at policy responses to address the challenge, beyond just monitoring, reporting and risk assessment (see chapter 2). Ongoing national policy responses and country case studies are documented, and categorised as: i) source-directed approaches (section 3.3), ii) use-orientated approaches (section 3.4), and iii) end-of-pipe measures (section 3.5) to managing pharmaceuticals in the environment.

3.3. Source-directed approaches

Source-directed policy approaches include different types of policy instruments, which impose, incentivise, or encourage measures in order to prevent the release of harmful substances to water bodies and their adverse effects on aquatic ecosystems. They are primarily targeted towards pharmaceutical companies and manufacturing facilities, however regulatory bodies, companies that purchase bulk drugs, international investment companies, pharmacies and healthcare organisations that buy and distribute final pharmaceutical products have the potential to influence and create incentives to manage their release (Larsson, 2014[1]). Table 3.1 summarises various source-directed regulatory, economic and voluntary policy instruments, which concern diverse, intersecting policy sectors, such as health, water, agriculture, industry and environmental protection. The following sections describe some of these policy instruments in more detail and provide case study examples of their implementation.

Source-directed approaches have the added advantages of reducing the need for, and cost of, end-of-pipe wastewater treatment upgrades and solid waste disposal, and supporting drinking water safety. Examples include green pharmacy, good manufacturing practices, EQNs and water quality standards, and water safety planning. Substance bans are another source-directed policy approach. For instance, diclofenac has been banned for veterinary use in India, Nepal and Pakistan as a result of the decline of vulture populations in the Indian subcontinent (Box 1.5). The German Environment Agency proposes that veterinary pharmaceuticals containing active substances with PBT or vPvB properties should be banned to protect the environment (UBA, 2018[2]). Where a substance ban is in the political discussion, one has to consider that such a restriction comes at the cost of employing them for what may be valuable human and animal health benefits, in comparison to the cost of damages if use is continued. It also requires consideration (alternatives assessment) that a substitute pharmaceutical is available with lower environmental risk (i.e. does not simply result in pollution-swapping).

Table 3.1. Source-directed policy instruments to prevent the release of pharmaceutical residues to water bodies

Policy instrument type	Policy instrument	Description
Regulatory	Substance bans	Complete prohibition of non-essential use of problematic pharmaceuticals
	Marketing authorisation	Evaluation-dependent authorisation of pharmaceuticals based on their predicted risks to human health and the environment. Such evaluations can also take into account principles of green chemistry such as "rational design" or "benign by design". Incentives can be used for green pharmaceuticals, such as fast-track marketing authorisation, reimbursement for greener APIs or longer exclusivity.
	Environmental quality norms and water quality standards	EQNs and water quality standards for harmful substances in water bodies. Detection above safe levels (or PNEC) can require action upstream to protect water bodies from harmful effects.
	Green public procurement	Clear and shared environmental criteria (and performance indicators) to pre-qualify pharmaceuticals for public procurement. Added advantage of impacting trade of pharmaceutical products across country borders.
	Good manufacturing practices and audits	Mandatory codes of conduct to reduce emissions from pharmaceutical manufacturing plants, as part of good manufacturing practice audits. Alternatively, environmental criteria for green public procurement could incorporate good manufacturing practices.
	Effluent discharge permits	Effluent discharge permits issued to pharmaceutical manufacturing plants with conditions for protection of drinking water sources and freshwater ecosystems. Non-compliance may lead to fines or withdrawal of operation permits.
	Best available techniques	Guidance documents that assist industrial operators with the design, operation, maintenance and decommission of manufacturing plants in compliance with environmental quality standards and discharge permit conditions (i.e. based on the PNEC or safe level of discharge). A BAT-based approach can be used to help set emission limit values as part of discharge permit conditions.
	Drinking water quality standards and water safety planning	Preventive measures to identify and address the source of risks to drinking water.
Economic	Subsidies for "green" action	Financial support from governments in return for environmental commitments by the private sector, such as reduced pollution from pharmaceutical manufacturing facilities.
	Subsidies for green pharmacy innovation	Subsidies or tax incentives for innovations green pharmacy, biological therapies, personalised or precision medicines to improve the business case for industry.
	Pollution charges	Charges to pharmaceutical manufacturing plants for discharging toxic effluent to water bodies.
Voluntary	Information campaigns	Transfer of knowledge or persuasive reasoning to industry on how to avoid water pollution.
	Voluntary agreements between private and public sectors	Non-legally binding agreements negotiated on a case-by-case basis between industry and public authorities fixing environmental targets or specific mitigation measures (e.g. changes in the production chain).

Source: Author

3.3.1. Environmental quality norms and water quality standards

It remains to be seen if environment quality norms (EQNs; also commonly known as environmental quality standards) are a feasible option to address pharmaceuticals given the number of APIs and the time taken to develop quality norms. Existing EQNs aim to control residual compounds, but there is a need to account for the impact of a mixture of pharmaceuticals and chemicals that ecosystems and humans are exposed to. There can also be a long time lag between the identification of a substance as having potentially negative impacts and the introduction of the associated EQN in legislation. As a consequence, the current control management of pharmaceutical pollution is often reactive i.e., it is in response to problems with water quality. This is demonstrated through the case of the detection of pyrazole in drinking water and the development of an EQN in the Netherlands (Box 3.1). The Netherlands has recognised the limitations of such an approach and have subsequently established other precautionary measures, such as: i) issuing wastewater discharge permits conditional to the protection of drinking water sources from pharmaceuticals and other emerging pollutants, ii) formation and use of a protocol for risk-based monitoring of pharmaceuticals and other emerging pollutants for drinking water quality by all drinking water companies (Box 3.1), and iii) improved stakeholder engagement and action across the pharmaceutical life cycle (section 4.2).

In 2012, the European Commission proposed the setting of EQNs for four pharmaceuticals of emerging concern: diclofenac, estrone (E1), 17-β -estradiol (E2) and 17-α-ethinylestradiol (EE2). In the UK, estimates made in 2013 by the water industry and regulators tentatively put the cost of "end of pipe" treatment for these chemicals at GBP 27-31 billion over a 20-year period (see section 4.4.4). Ultimately, the evidence base for regulating these pharmaceuticals in the EU was felt to not be strong enough in the light of potential costs; they were however placed on the WFD's "watch list" for ongoing monitoring (see Box 2.4).

Box 3.1. Deriving a water quality standard for Pyrazole, Netherlands

In the Netherlands, drinking water suppliers and national government agencies control drinking water quality by monitoring different parameters, either enforced by legislation or as part of non-binding efforts. In addition to this, non-target screenings are performed with the aim to identify new potential hazardous substances in drinking water supplies. A threshold value (or signalling parameter) of 1 µg/L is used to identify unregulated substances in drinking or groundwater. Exceedance of this threshold value does not indicate an immediate threat to humans, but triggers the investigation of possible human health effects.

In 2015, this threshold value was exceeded in the River Meuse (an important drinking water source) by an unknown compound (concentration detected at 100 µg/L). This compound was later identified as Pyrazole. The cause of the elevated concentrations of Pyrazole was due to a malfunctioning WWTP. At the time, little was known about the human health impact of pyrazole. Given the potential threat to drinking water quality, it was decided that precautionary action should be taken. As a consequence, three drinking water companies were required to temporarily use alternative water sources for drinking water supply.

The incident triggered the development of a water quality standard (WQS) for Pyrazole, implemented in the national Dutch Drinking Water Directive on July 2017 (IENM/BSK-2017/160338). Toxicity testing and existing data were used to help determine the WQS. In addition, the following four factors were taken into account when determining the standard: i) the ability of treatment systems in place to remove pyrazole; ii) the precautionary principle to take into account possible risks where scientific understanding was not yet complete, including for mixture toxicity; iii) maintaining consumer's trust in drinking water; and iv) water quality standards from neighbouring transboundary country Germany.

The Pyrazole incident also initiated The Structural Approach Programme – a programme designed to improve the long-term risk prevention of CECs in drinking water. The programme was developed by the Ministry of Infrastructure and Water Management, together with relevant stakeholders from regional and local government agencies, industry, and knowledge institutes. Taking into account lessons learned in the Pryazole case, stakeholders recommended the following points for improvement:

- The implementation of a systematic assessment of the potential effects of the emission of anthropogenic substances, specifically on drinking water quality, in the issuing process of wastewater discharge/emission permits. In response, the relevant wastewater guidelines was modified to mandate the inclusion of the potential effects of CECs from effluent on drinking water production.

> - The formation and use of a protocol for risk-based monitoring of drinking water quality by all drinking water companies by 2019. It is anticipated that this recommendation will be implemented in federal law.
> - The formalisation of the procedure for setting drinking water quality values and making them available to all stakeholders.
>
> At the same time, a step-by-step-guide was created with the purpose of improving the management of incidents of CECs in drinking water in the short-term.
>
> The guide has been designed to improve information exchange between stakeholders at an early stage of an incident. These policy responses have increased the level of cooperation between, and commitment amongst, stakeholders, and created a mutual understanding as well as mandatory requirements.
>
> *Source*: Summary of case study provided Julia Hartmann, Ana Versteegh and Tialling Vlieg, Ministry of Infrastructure and the Environment, Netherlands.

3.3.2. *Green pharmacy*

Ultimately the use of pharmaceuticals which are biodegradable and have a low intrinsic toxicity would be the most 'eco-effective' and sustainable in a circular economy. However, potency and stability are often important properties of effective pharmaceuticals for human and animal health. A potential long-term approach to reducing the environmental risks of pharmaceuticals, and possible health risks at source, is the rational design and manufacture of new green pharmaceuticals (or 'benign by design' (Kümmerer, 2007[3])). 'Green pharmacy', 'green chemistry' or 'sustainable chemistry' is often referred to as: i) the development of new substances that are more efficiently biodegraded but retain their effective pharmaceutical properties (Van Wezel et al., 2017[4]; Anastas and Warner, 1998[5]) (Leder, Rastogi and Kümmerer, 2015[6]), or ii) the re-design of existing pharmaceuticals for environmental biodegradability (Rastogi, Leder and Kümmerer, 2015[7]) (although a definition is not agreed upon). The expected outcome is better biodegradable and pharmacologically active drug molecules that do not accumulate in, or cause adverse effects to, the environment. An example is the development of glufosfamide - a green alternative to ifosfamide (chemotherapy medication) - that is 70% biodegradeable and has the added benefit of improved uptake in the human gut (Kümmerer, 2007[3]; Daughton, 2003[8]). Biologics - any pharmaceutical drug product manufactured in, extracted from, or semi-synthesised from biological sources – also have shorter half-lives.

Although research on green pharmacy has expanded in recent years, the share of green pharmaceuticals on the market is still low. Green pharmacy is expected to bring positive environmental results in the medium to long-term. The 2018 Nobel Prize in Chemistry, for example, was awarded for path-breaking research on how chemists produce new enzymes, leading to new pharmaceuticals and cancer treatments and less waste (UN Environment, 2019[9]). A new generation of biodegradable antibiotics (which would rapidly degrade in WWTPs) together with targeted delivery mechanisms, could reduce accumulation in freshwater ecosystems and subsequently to the promotion of resistance development.

Barriers delaying immediate progress of green pharmacy include (Matus et al., 2012[10]):

- *An agreed definition.* An agreed definition of green pharmacy is required between environmental chemists, clinical chemists, drug discovery scientists and other relevant stakeholders. The definition must be realistic and risk-based, with the aim of achieving a balance between patient/animal health and environmental protection.
- *Regulatory barriers.* Pharmaceutical manufacturers have little incentive to invest in pollution prevention and green pharmacy; most environmental, health, and safety regulations focus instead on controlling risk through reductions in exposure with end-of-pipe technologies. For example, in

2005 in the U.S., chemical manufacturing spent more than any other industrial sector on pollution abatement (U.S. Census Bureau, 2008[11]), a partial reflection of a regulatory focus on risk control, rather than risk prevention.

- *Economic and financial barriers*: costs, incentives, and markets. A green pharmaceutical must not only represent an improvement for health and the environment, but it must also be profitable, without sacrificing efficacy or quality. In the case of an existing pharmaceutical, changes in the formulation must represent enough of a potential cost savings to outweigh upfront costs.

- *Technical barriers*: uncertainties, expertise and metrics. The science behind green chemistry is often complex and multidisciplinary, and there are significant uncertainties. The novel field of green pharmacy requires combinations of innovative methods, and linkages between environmental science and green chemistry, with other existing pharmaceutical research fields in pharmacology, drug development, organic chemistry, computational chemistry and analytical chemistry (Leder, Rastogi and Kümmerer, 2015[6]). There are many reactions and processes for which greener substitutes remain unknown. New formulations and degradation products should be tested with the same rigor (and vigor) as conventional products for toxicity; parent compound toxicity must be considered, as well as the toxicity of degradation products. An absence of clear definitions and metrics for use by researchers and decision makers also pose barriers. Finally, advances in green pharmacy may not be shared by pharmaceutical companies in order to retain their competitive advantage.

Policy can help to reduce these barriers and provide incentives to make green pharmacy more attractive to pharmaceutical companies. For instance, governments can:

- Facilitate knowledge creation and accessibility, through R&D spending and access to data (including testing) to avoid costly replication of work.

- Demonstrate the feasibility of new green pharmaceuticals as new business opportunities for industry.

- Create tax incentives, access to inexpensive capital and technical assistance for implementation to drive innovation (OECD, 2012[12]). A return on public investments in new pharmaceuticals should be considered when assessing subsidies for the private sector in pharmaceutical development.

- Implement ERA-dependent authorisation of new pharmaceuticals based on their predicted risks to the environment to incentivise investment in green pharmacy. Such evaluations could take into account principles of green chemistry such as "rational design" or "benign by design", and allow for an easy or fast-track authorisation process based on the biodegradability of APIs and their transformation products after use. Other options for incentives for green pharmacy may include reimbursement for greener APIs or longer exclusivity.

- Develop and implement evidence-based technical guidelines on sustainable procurement of pharmaceuticals. Integrate environmental criteria into good manufacturing practices utilised by authorities to pre-qualify pharmaceuticals for procurement. There are initiatives to advance sustainable procurement of pharmaceuticals in order to create an incentive for manufacturers to strive towards production of more green products, as well as to integrate environmental criteria into manufacturing practices (SAICM, 2015[13]). For example, the interdisciplinary research project "PharmCycle" aims to reduce the contamination of the aquatic environment with antibiotics by developing sustainable antibiotics, improving the ERA of antibiotics (and increasing data availability), and reducing the discharge of antibiotics in wastewater (Andrä et al., 2018[14]).

3.3.3. *Water safety planning*

As outlined in the WHO 2011 *Guidelines for Drinking Water Quality*, the water safety plan (WSP) approach is "the most effective means of consistently ensuring the safety of a drinking-water supply [...] through the use of a comprehensive risk assessment and risk management approach that encompasses all steps in the water supply from catchment to consumer". Water safety plans highlight the importance of considering risk assessment and risk management comprehensively from source to tap, and adopting preventive measures to address the source of risks (WHO, 2012[15]).

The primary objectives of a WSP in ensuring good drinking-water supply practice are: i) the prevention or minimisation of contamination of source waters; ii) the reduction or removal of contamination through treatment processes; and iii) the prevention of contamination during storage, distribution and handling of drinking-water. For more information on water safety planning, refer to the WHO Water Safety Plan Manual (Bartram et al., 2009[16]).

Adapting the water safety plan approach to the context of pharmaceuticals in drinking water means that preventing pharmaceuticals from entering the water supply cycle during their production, consumption (i.e. excretion) and disposal is a pragmatic and effective means of risk management. Preventive measures need to be applied as close as possible to the source of the risk and hazard (WHO, 2012[15]).

3.3.4. *Good manufacturing practices, best available techniques and green public procurement*

Good manufacturing practices (GMP) and Best Available Techniques (BAT) are one policy tool to prevent and control the emission of industrial pollutants, and thus to improve the protection of human health and the environment (OECD, 2018[17]).

BAT are guidance documents that assist industrial operators with the design, operation, maintenance and decommission of manufacturing plants in compliance with environmental quality standards and discharge permit conditions (which are legally binding) to prevent or control emissions to air, soil and water. A BAT-based approach can also be used to help set emission limit values as part of discharge permit conditions. One of the advantages of BAT is that it gives industrial facilities the freedom to choose their preferred means to achieve environmental quality standards and discharge permit conditions, using BAT as a guiding – rather than prescriptive – policy tool.

Determination of BAT is often done in consultation with stakeholders (OECD, 2018[17]). Technical and environmental aspects are taken into account as part of the evaluation of techniques, and in most cases also economic aspects. The EU operates with a hierarchy of techniques, giving priority to preventive techniques, over end-of-pipe measures when selecting BAT. The EU and the U.S. both have BAT for pollution prevention and control during the manufacturing of pharmaceutical products (EU, 2010[18]) (EC, 2006[19]; US EPA, 2006[20]).

GMP focus on quality assurance to ensure that medicinal products are consistently produced and controlled to the quality standards appropriate to their intended use and as required by the product specification (WHO, 2014[21]). Although GMP do not currently focus on environmental protection, there is potential to integrate environmental criteria, such as effluent discharge standards, into GMP utilised by the WHO to pre-qualify pharmaceuticals for procurement. Because any company that exports pharmaceuticals must follow GMP, such criteria have the potential for a wider impact (Larsson, 2014[1]). The process of incorporating environmental criteria into GMP would need to be carefully managed to ensure buy-in and prevent withdrawal of GMP agreements by countries.

GMP, sustainability reporting and green public procurement may be particularly relevant to OECD pharmaceutical companies operating in developing economies. For example, in Sweden, about one-third of antibiotics on the market are produced in other regions where the environmental requirements and enforcement are in general weaker (e.g. India, China and Puerto Rico) (Bengtsson-Palme, Gunnarsson and Larsson, 2018[22]). The Swedish government has proposed a revised system in which pollution control during manufacturing is considered when companies compete to obtain product subsidies for state healthcare. Swedish county councils have also started to request monitoring of emissions during manufacturing when procuring medicines (Larsson, 2014[1]). Environmental criteria and performance indicators to pre-qualify pharmaceuticals for public procurement has the added advantage of impacting trade of pharmaceutical products across country borders.

Connecting control of emissions to the existing regulatory framework of GMP would allow for control of the production chain, thus taking advantage of a system of control and action that is already in place. Additional benefits include preserving a level playing field regarding effects on competition between companies or countries within and outside Europe. Sweden have recently proposed that within the EU GMP framework, requirements limiting emissions of APIs into the water environment be included in the EU Directives on Medicinal Products for Human Use and Medicinal Products for Veterinary Use (Swedish Medical Products Agency, 2018[23]).

3.4. Use-orientated approaches

Use-orientated policy approaches include policy instruments which impose, incentivise, or encourage a reduction in the use of pharmaceuticals and their release to the environment. They are designed to inform and change the behaviours and practices of physicians, veterinarians, pharmacists, patients and farmers. For example, a reduction of inappropriate and excessive consumption of pharmaceuticals can be achieved through: i) regulatory instruments, such as restrictions on over-the-counter sale of environmentally harmful pharmaceuticals, and mandatory codes of conduct for health practitioners, veterinarians and farmers; ii) economic instruments, such as subsidies from governments in return for environmental commitments by the private sector; and iii) voluntary measures, such as public environmental health campaigns on the sustainable use and disposal of pharmaceuticals.

Use-orientated policy instruments for pharmaceuticals have not long been considered; in general they involve the modification of long-established norms in the practice of clinical prescribing, and human and animal health practices. However, as excretion and release through WWTPs is one of the most important routes of pharmaceuticals into the environment, use-orientated approaches have the potential of being more effective and less-costly in comparison to end-of-pipe solutions (Daughton, 2014[24]). There are opportunities for alternative business models in the health care and agriculture sectors based on increased service and education, instead of resorting to pharmaceuticals only.

Table 3.2 summarises various use-orientated-directed policy instruments, most of which are aimed at health practitioners, pharmacists, veterinarians, farmers and the general public. The following sections describe some of these policy instruments, targeting health-care practices, and veterinary and agriculture practices. A number of case studies are provided as examples of policy implementation.

Table 3.2. Use-orientated policy instruments to reduce release of pharmaceutical residues to water bodies

Policy instrument type	Policy instrument	Description
Regulatory	Substance bans	Prohibition of non-essential use of pharmaceuticals
	Substance restrictions	Restrictions on the prescription of non-essential use of pharmaceuticals
		Restrictions on over-the-counter sale and purchase of environmentally harmful pharmaceuticals
		Constraints to the placement on the market or the use of a substance at specific points in time (e.g. before rainfall events) or locations (e.g. sensitive areas). This is particularly relevant to farming practices.
	Best environmental practices (BEP) for health care practices	Mandatory codes of conduct for health practitioners to promote sustainable use (e.g. improved diagnostics, rationale use and targeted drug regiments) and responsible disposal of pharmaceuticals
	Best environmental practices (BEP) for veterinary and agriculture practices	Mandatory codes of conduct for veterinarians and farmers to promote improved diagnostics, sustainable use and disposal of pharmaceuticals, and reduce emissions from veterinary and agriculture practices
Economic	Product charges	Tax levied on products high-risk APIs in order to incentivise consumers to reduce or change consumption behaviours. Pharmaceuticals that are of high-risk to the environment and difficult to remove with conventional wastewater treatment could be priced accordingly.
	Substance charges	Tax levied on hazardous compounds in order to incentivise producers to change production processes or substitute chemicals with less hazardous alternatives (e.g. green pharmacy)
	Subsidies for "green" action	Financial support from governments in return for environmental commitments by the private sector
Voluntary	Public environmental health campaigns and disease prevention	Transfer of knowledge or persuasive reasoning on sustainable use, consumption and disposal of pharmaceuticals, and how to prevent illness and the need for pharmaceuticals (e.g. through effective hand washing to prevent spread of infection)
	Eco-labelling of green pharmaceuticals	Products that meet certain environmental standards can be marketed and sold at a premium and/or subsidised (n.b. regulatory changes would be required to set up such schemes)

Source: Author

3.4.1. Health-care practices and consumer choices

It is explicitly understood that pharmaceuticals exert an indispensable factor in modern healthcare and that the general public should have access to the best available pharmaceuticals. The challenge is to establish pharmaceutical management that combines the best possible treatment of the patient with a sustainable environmental cautiousness.

Health practitioners and pharmacists play an important role in influencing and changing citizen's behaviour on the sustainable use and disposal of pharmaceuticals. One of the most important aspects is to stress the importance of avoiding prescription of unnecessary medicines. Alterations to prescribing and dispensing practices may include:

- Avoiding prescription of unnecessary medicines with improved diagnostics and increased service and education focusing on prevention rather than cure (e.g. hand-washing, diet/nutrition, well-being and therapy). Antibiotics, for example, should be prescribed only when there is an evidence-based need in order to reduce the risk of resistant strains. A successful response to AMR will address not only antimicrobials and the over- and mis-use of antibiotics, but also diagnostics, vaccines and alternatives to antibiotics for human and animal health (WHO, 2015[25]).

- Lower, optimised and more targeted dose prescribing (including shorter regimens and refill schedules) tailored to patients needs can minimise release of pharmaceuticals to the environment.

Personalised adjustment of drug dose holds the potential for also enhancing therapeutic outcomes while simultaneously reducing the incidence of drug side-effects, optimising use of resources and lowering patient healthcare costs (Daughton and Ruhoy, 2013[26]). Genomics enables patients to receive therapy individually customised to their genetic makeup. Much of the cost of DNA sequencing is dropping, and in the future, may change the whole concept of prescribing medication (UN Environment, 2019[9]).

- Prescribing smaller packages, to deliver the exact dosage of medicine required, rather than standard universal packaging. However, this may increase costs for pharmacists.
- Substitution to green pharmaceuticals with less environmental impact but equivalent medicinal benefits (see section 3.3.2).

Education and information campaigns play a major role in the adoption of use-orientated approaches. Environmental issues can be introduced in already existing information schemes to increase awareness that, in addition to the desired health effects, certain pharmaceuticals may have significant environmental impacts. In Germany, a two-year training and education campaign on the sustainable consumption and responsible disposal of pharmaceuticals in the town of Dülmen (population circa 46,000) was particularly effective. Results from the campaign - involving 13 schools, all pharmacists, and many doctors, sports clubs and local stakeholders – increased awareness of the environmental issues to 77% of the population, increased appropriate disposal of unused pharmaceuticals by 20%, and reduced the number of people using painkillers between medical treatments by 10% (noPILLS, 2015[27]). Information received via trustworthy avenues, such as the doctors' waiting room and the annual pharmaceutical garbage collection calendar were particularly effective (noPILLS, 2015[27]). Health insurers are another stakeholder that should be involved in the discussion. They may play a role in the reimbursement of environmentally friendly alternatives such as health education, non-medicine treatments and green pharmaceuticals.

Action on pharmaceuticals in the environment is much more likely to be extended and sustained if it is mainstreamed into broader health, agricultural and environmental projects. The WHO One Health approach provides such a framework, and recognises that putting resources into AMR containment now is one of the highest-yield investments a country can make to mitigate the impact of AMR (IACG, 2018[28]) (WHO, 2015[25]). Providing access to safe clean drinking water and sanitation services, in line with Sustainable Development Goal 6, will be critical to reducing AMR, and improving human health and well-being. Given the scale of the AMR crisis (Box 1.7, chapter 1), the OECD (2018[29]) has undertaken analysis of five simple policy interventions to reduce AMR targeting different stakeholders. The analysis has demonstrated their high impact on population health, affordability to implement and favourable cost-effectiveness ratio (Box 3.2). Furthermore, when such policies are implemented together as part of a coherent strategy, they produce a more beneficial impact.

Box 3.2. OECD cost-effective policy actions to combat antimicrobial resistance

OECD analysis demonstrates that governments could counter the AMR problem with five affordable policies as part of a coherent package. The policies align with the WHO's Global Action Plan on AMR.

1. Improve hygiene in healthcare facilities, including promotion of hand-washing and better hospital hygiene.

2. Introduce stewardship programmes promoting more prudent use of antibiotics to end their over-prescription and over-use.

3. Utilise rapid diagnostic tests to detect whether an infection is bacterial or viral before issuing prescription antibiotics.

4. Delay prescription of antibiotics when they are not immediately required.

5. Establish public awareness campaigns on the impacts of AMR and the importance of: hand-washing, completing a full treatment of antibiotics as prescribed (to avoid re-infection) and the safe disposal of unused antibiotics to avoid their entry into the environment.

It is projected that investment in these measures could pay for themselves within one year and produce savings of about USD 1.5 for every dollar invested thereafter. Simple measures, such as promoting hand washing and better hygiene in healthcare facilities, more than halve the risk of death and decrease the health burden of AMR (measured in DALYs[1]) by about 40%. Mass media campaigns, delayed prescriptions and the use of rapid diagnostic tests also produce a positive health impact, albeit more limited.

Public health actions to tackle AMR are affordable. Implementing such policies varies from as little as USD PPP[2] 0.3 per capita for mass media campaigns to a few hundred USD PPP per hospitalised patient in the case of enhanced hygiene in healthcare.

1: DALYs = Disability Adjusted Life Years. DALYs is the sum of years of potential life lost due to premature mortality and the years of productive life lost due to disability.
2: PPP = Purchasing power parity.
Source: (OECD, 2018[29]).

Classification and labelling approaches may help to minimise risks. A good example is found in Stockholm, Sweden, where pharmaceuticals are classified according to their environmental risks and this information is distributed to doctors and made publicly available to facilitate the prescription and use of pharmaceuticals with lower environmental risk (Box 3.3). The extension of a model similar to the Swedish scheme could potentially be desirable on a European level or to other OECD countries. Key issues for developing and implementing classification and labelling schemes include: necessary improvements to ERAs (see section 2.2), and the standardisation of the information used, the criteria applied, who provides the information and the mode of communication (Clark et al., 2008). It also requires careful consideration (an alternatives assessment) that a substitute pharmaceutical is available with lower environmental risk (i.e. does not simply result in pollution-swapping).

Box 3.3. Environmentally Classified Pharmaceuticals: Allowing doctors to make informed prescription choices, Sweden

Reducing pharmaceuticals in water is an environmental priority of the Stockholm County Council. Essential Drug Recommendations were issued and launched as a 'Wise List' by the regional Drug and Therapeutics Committee in Stockholm, on the basis of environmental hazard assessments initiated in 2003 by the Environmental Department of the Stockholm County Council. The Wise List is a list of recommended pharmaceuticals for the treatment of common diseases that takes into account cost-effectiveness and environmental impact when comparing medications that are equally safe and equally suitable for the medical purpose.

The acute risk to the aquatic environment was one of the assessment criteria of pharmaceuticals on the Wise List, with pharmaceuticals scored as insignificant, low, moderate or high risk according to environmental persistence, bioaccumulation and toxicity. These scores are used to assist with decision-making by doctors when prescribing medications to their patients. In 2009, 77% of doctors in Stockholm were reported to have adhered to the Essential Drug Recommendations and the Wise List.

A 2014-2015 leaflet of Environmentally Classified Pharmaceuticals was issued to all prescribers with the following information: i) a brief summary of the impact of pharmaceuticals on the environment, ii) the 'Wise List', including details on how the substances are classified according to their environmental risk and how to interpret the classification of pharmaceuticals (including which pharmaceuticals are exempt), iii) the role of the precautionary principle, and iv) advice for doctors and patients to reduce environmental impacts.

Sources: (Gustafsson et al., 2011[30]; Stockholm County Council, 2014[31]).

Pharmacists can also play an important role in educating the public on the environmental effects of pharmaceuticals. For example, Swedish pharmacies are required to educate consumers on the environmental effects of diclofenac, for which there is a water quality standard governed by the Marine and Water Authority's regulations on specific pollutants (Box 3.4). Viable, 'greener' substitutes for diclofenac is something that is questioned; there is concern of pollution swapping with substitutes such as naproxen and ibuprofen, which may have similar environmental impacts to diclofenac (limited data).

Box 3.4. Swedish pharmacies required to educate consumers on the environmental effects of Diclofenac

In 2015, Sweden introduced legislation on how much diclofenac may be released into water, under the Marine and Water Authority's regulations on specific pollutants. This value was often exceeded. Diclofenac is a popular analgesic drug but can have negative environmental impacts (see section 1.4) which Sweden believes customers should be aware of.

In 2018, the Swedish Association of Pharmacy agreed that all pharmacies must start to inform consumers about diclofenac's adverse effects on the environment. Shelf signs inform consumers that the substance has a negative impact on the environment and should be used with reflection. The agreement applies to both store and e-commerce sales.

Many Swedish county councils have removed diclofenac from their recommended lists of prescription drugs because of its environmental impact. In addition, the Swedish Medical Products Agency, which authorises pharmaceuticals in Sweden wants a regulatory change requiring environmental effects be considered in the authorisation and approval of human pharmaceuticals.

Swedish pharmacies also take other steps to reduce pharmaceuticals in the environment. For example, in 2017, Swedish pharmacies collected 1 200 tonnes of unused drugs as part of a public collection scheme. Several pharmacies have introduced their own eco-labels for over-the-counter pharmaceuticals.

Sources: (Swedish Association of Pharmacy, 2018[32]) (Ringbom, 2017[33]) (Sweco, 2016[34]).

3.4.2. *Veterinary and agricultural practices*

The overuse of veterinary pharmaceuticals, such as antibiotics and hormones used as growth promoters in industrial farming, results in the release of their residues into soil, groundwater and surface waters via leakage from animal waste storage and disposal tanks, and the use of animal manure and slurry as fertiliser and irrigation water. Agriculture is not only a source of pharmaceutical pollutants, but also contributes to the spread and introduction of these pollutants into aquatic environments through municipal wastewater reuse for irrigation and the application of biosolids onto the land as fertiliser (Muñoz et al., 2009[35]). Use of veterinary pharmaceuticals in aquaculture directly enter water bodies

Regulations specifically addressing these sources of pharmaceutical pollutants are lacking at national and international levels. The overuse and misuse of antibiotics and parasiticides (e.g. as a preventive measure) and hormones (e.g. as growth promoters) of are of particular concern because of their contribution to AMR (see Box 1.5), their PBT properties, and endocrine disrupting effects (see Box 1.9), respectively.

Changes in agricultural practices may minimise the risks of pharmaceuticals in the environment. A range of approaches can be used, including (Pope et al., 2008[36]; Boxall, 2012[37]):

- Minimising the use of veterinary medicinal products
- Preventive measures for improving animal health
- Requiring more targeted treatment of sick animals, rather than treating a whole herd
- Changes in treatment timings and intensities
- Development of best practices on manure storage and treatment to increase biodegradation of APIs before land application
- Changes in manure/sludge application rates and timings (i.e. to avoid/minimise overland runoff and leaching). For example, injection application of manure has been shown to reduce overland runoff of pharmaceuticals when compared to a broadcast application (Topp et al., 2008[38]). And the application of sewage sludge during dry periods would minimise the potential for some substances to be transported to surface waters (Boxall, 2012[37]).
- Development of recommendations on where not to apply manure and biosolids (e.g. where slopes or soil types are unsuitable) and
- Specification of riparian buffer zones to protect water bodies.

Recently, several effective and alternative means of treating and controlling livestock disease caused by microorganisms have been developed. This includes new gene editing technology; effective immunoglobulins, which can act in the body as defence molecules against infections; and peptides, which help the host system in its defence against pathogens, partly by delaying the establishment of infection. The production of peptides can be induced in the host animal, or can be used to create vaccines, and thereby prevent illness and reduce the need for antibiotics (Marquardt and Li, 2018[39]; van Dijk et al., 2018[40]). In Norway, the development of a vaccine to prevent furunculosis (a highly-contagious bacterial fish disease) has significantly reduced the need for antibiotics in salmon farming (WHO, 2015[41]). Improved DNA technology has also greatly facilitated the selection of livestock that have genetic resistance to pathogenic microorganisms (Marquardt and Li, 2018[39]).

The use of veterinary pharmaceuticals can be optimised and reduced through information campaigns. For example, Germany has developed an environmental checklist for the use of pharmaceuticals in veterinary medicine and livestock farming (Box 3.5).

Box 3.5. Environmental checklist for the use of veterinary medicines

In 2017, the German Environment Agency (UBA) developed an environmental checklist for the use of veterinary pharmaceuticals, with the aim of reducing pharmaceutical residues in the environment. The checklist is targeted to veterinarians and farmers. A summary translation of the checklist is outlined below.

Veterinary Medicine

- Operate sufficient animal health diagnostics
- Confirm a medical indication
- Consider whether avoiding drug treatment is possible
- Consider alternative remedies to avoid input of pharmaceuticals to the environment, and to reduce resistance formation (i.e. contribution to AMR)
- Inform the pet owner about the correct administration of the prescribed medication
- Inform the pet owner about the correct storage and disposal of the veterinary drug.

Livestock Farming

- Question whether preventative measures, such as vaccinations, are necessary
- Ensure the veterinarian and farmer have considered alternative treatment methods
- Follow the veterinarian instructions regarding administration of the correct dosage, duration and frequency of the medicine.
- Keep a record of veterinary medicines to avoid unnecessary treatment and entry of pharmaceutical restudies to the environment
- Dispose of used medicines, leftovers and packaging properly
- Ensure a rest period of several months before the spreading of manure, which may contain traces of veterinary medicinal products
- Assess what prevention measures could be taken at the farm scale to avoid or reduce future drug treatments.

Source: (Kemper, Hein and Lehmann, 2017[42]).

Restrictions on the unnecessary use of veterinary pharmaceuticals have proven to be effective. Since 2006, antibiotics used as growth promoters in feed additives within the EU have been banned (European Commission, 2005[43]). In addition, veterinary prescriptions are required to use antibiotics (i.e. they cannot be purchased over-the-counter, with exemptions in certain cases). Implementation of these EU policies has resulted in a decreased detection of antibiotic resistant genes and pathogens in Germany, the Netherlands, Sweden and Denmark. In Germany, since the introduction of the German Antimicrobial Resistance Strategy and the 16th amendment to the German Medicines Act, the consumption of antibiotics for livestock farming has reduced by over 32% between 2014 and 2017 (BMEL, 2019[44]). In the UK, the poultry industry has successfully reduced unnecessary antibiotic use – whilst increasing meat production – with a voluntary antibiotic stewardship programme (Box 3.6). Canada has implemented a similar policy to the EU, requiring that antibiotics for animal use be sold by prescription only. Further actions to promote responsible use in Canada include: i) removal of growth promotion claims from pharmaceutical labels; and ii) labelling all antibiotics to be used as additives in livestock feed and water with responsible use statements (Government of Canada, 2018[45]).

Box 3.6. Voluntary reductions in the use of veterinary antibiotics: The British Poultry Council Antibiotic Stewardship

In the UK, the British Poultry Council Antibiotic Stewardship aims to ensure the sustainable use of antibiotics throughout the supply chain, based on three measures: i) review and replace antibiotics used where effective alternatives are available; ii) reduce the number of birds receiving treatment, through systems based on risk assessments; and iii) continue to refine existing strategies, using data collection. More specifically, it was agreed to: i) cease preventative use of antibiotics; ii) restrict the use of antibiotics classified as critically important by the WHO; and iii) ban the use of third and fourth generation cephalosporins (a class of antibiotics).

As a result of the Stewardship, over the period 2012-2017, total antibiotic use in the British poultry industry reduced by 82% and the use of Fluoroquinolones (a Critically Important Antibiotic for human health) reduced by 91%. During the same period, poultry meat production increased by 10 %.

Source: (British Poultry Council, 2018[46]).

In Denmark, strict biosecurity measures and the Specific Pathogen Free system have kept the country free from many pig diseases and helped contain infected animals so that the spread between farms is limited (FAO and Denmark Ministry of Environment and Food, 2019[47]). The focus on keeping diseases out of farms is the most important step to lower antimicrobial consumption. The phasing out of antimicrobial growth promoters in the year 2000 in Denmark has showed that it is feasible to reduce the use of antimicrobials for pigs, and sustain low usage, whilst: i) maintaining high productivity; ii) resulting in no detrimental effects on pig health or welfare; and iii) and at low cost to the farmer. One of the reasons for Denmark's success was the stepwise phasing out of antimicrobial growth promoters over several years that gave farmers sufficient time to adjust (FAO and Denmark Ministry of Environment and Food, 2019[47]).

3.5. End-of-pipe measures

End-of-pipe measures focus on removing or eliminating pharmaceuticals after their use or release into water. End-of-pipe policies involve different types of instruments that impose, incentivise, or encourage improved wastewater treatment and solid waste disposal. One policy option for the reduction of pharmaceuticals in water is to provide incentives for the upgrade of WWTPs with new removal technologies; conventional WWTPs are not designed to, nor do they fully, remove pharmaceuticals. Another option involves improvement of solid waste collection and disposal to prevent unused pharmaceuticals from entering the water cycle.

End-of-pipe measures will not solely solve the problem of pharmaceuticals in water. They are limited by their removal efficiencies, high capital investment and operation costs and increased energy consumption. Furthermore, end-of-pipe measures do not capture diffuse sources of pharmaceutical pollution (e.g. from agriculture and aquaculture). However, end-of-pipe measures will remain a policy option due to the continued need of pharmaceuticals for human health and wellbeing.

Table 3.3 summarises end-of-pipe regulatory, economic and voluntary policy instruments, which concern a diverse set of stakeholders including consumers, wastewater utilities, solid waste collectors, pharmaceutical industry, and health and environment officials. The following sub-sections describe some of these policy instruments in more detail and provide case study examples of their implementation. Much of this section focuses on the removal of pharmaceutical residues through advanced wastewater treatment, the cost-efficiency of such a measure, and potential financing mechanisms to fund WWTP upgrades.

Table 3.3. End-of- pipe policy instruments to remove pharmaceuticals after their use and release into water bodies

Policy instrument type	Policy instrument	Description
Regulatory	Best available technique (BAT)	Definition of the best technology options for improved wastewater treatment
	Wastewater treatment standards	Definition of performance standards for wastewater treatment (e.g. treatment capacity or effluent load) without requiring a specific technology
	Pharmaceutical disposal requirements	Standards on correct waste disposal, e.g. mandatory consumer-level "take-back" programmes for unused pharmaceuticals
Economic	Effluent/ emission charges	Tax on discharging wastewater to water bodies, in order to incentivise emission reduction
	Wastewater tariffs or taxes for WWTP upgrades	Tariffs or taxes designed to signal the cost of wastewater treatment to remove pharmaceuticals to the public and consumers
	Subsidies for improved wastewater treatment	Financial support from governments to incentivise operators to invest in advanced wastewater treatment; or to promote research on improved wastewater treatment
	Extended producer responsibility (EPR) schemes	Instead of consumers being responsible for the cost of wastewater and waste management, producers, to some extent, become responsible for financing the end-of-life costs (wastewater treatment and solid waste disposal). In principle, companies can internalise these costs and are incentivised to produce pharmaceuticals more cost-efficiently and sustainably
Voluntary	Advisory services	Support from governments in the form of information, advice, and consultancy about improved wastewater treatment or solid waste management
	Voluntary agreements on wastewater treatment	Non-legally binding agreements negotiated on a case-by-case basis between wastewater treatment operators and a public authority to improve wastewater treatment practices
	Waste collection/ take-back schemes	Voluntary schemes designed to collect and appropriately dispose of unused pharmaceuticals, thereby reducing their release to water bodies.

Source: Author

3.5.1. Upgrading wastewater treatment plants with advanced removal technologies

Removal effectiveness and cost efficiency

Advanced wastewater treatment processes, such as reverse osmosis, ozonation, activated carbon, membranes and advanced oxidation technologies, can achieve higher removal rates for pharmaceuticals in comparison to conventional secondary wastewater treatment (activated sludge processes, or other forms of biological treatment such as biofiltration).

The removal efficiencies of the different wastewater treatment processes depend on the various physico-chemical properties of APIs (and their metabolites), such as hydrophobicity, reactivity, molecular size and charge and biodegradability. Table 3.4 summarises advanced wastewater treatment options, and their advantages and disadvantages. Activated carbon adsorption (with both **powered activated carbon** (PAC) and granular activated carbon (GAC)), ozonation and filtration by nanofiltration or reverse osmosis membranes, have been demonstrated to effectively remove most pharmaceuticals. Switzerland and Germany have upgraded some WWTPs with activated carbon adsorption and ozonation. However, filtration with nanofiltration or reverse osmosis are found to be more cost-intensive. Nevertheless, reverse osmosis membranes have been implemented in potable reuse schemes in the U.S., Singapore and Australia because of the additional benefit of reducing salinity and metal reduction (Rizzo et al., 2019[48]). Ozonation generates toxic transformation products that will need to (and can) be removed post-treatment (Völker et al., 2019[49]).

Table 3.4. Advantages and disadvantages of advanced wastewater treatment options to remove pharmaceuticals

Advanced treatment	Advantages	Drawbacks
UV light with hydrogen peroxide (UV / H$_2$O$_2$)	• Moderate-good pharmaceutical removal at lab/pilot scale • Also effective as disinfection process	• Formation of oxidation transformation products • No full-scale evidences on removal • Higher energy consumption compared to ozonation, specifically when high organic matter concentration acts as inner filter for UV radiation.
Photo-Fenton	• High pharmaceutical removal • Use of solar irradiation • Also effective as disinfection process	• Formation of oxidation transformation products • No full-scale evidences on pharmaceutical removal • At neutral pH 7 addition of chelating agents necessary. • Large space requirements for solar collectors
UV light with titanium dioxide (UV / TiO$_2$)	• High pharmaceutical removal • Use of solar irradiation • Also effective as disinfection process	• Low kinetics • Formation of oxidation transformation products • Catalyst removal • Large space requirements for solar collectors
Ozonation	• High pharmaceutical removal • Full scale evidence on practicability • Partial disinfection • Lower energy demand compared to UV/H$_2$O$_2$ and membranes	• Formation of by-products and other unknown oxidation transformation products • Need for a subsequent biological treatment (e.g., slow sand filtration) to remove organic by-products
Powdered activated carbon (PAC)	• High pharmaceutical removal • Full scale evidence on practicability • Additional dissolved organic carbon removal • No formation of by-products • Partial disinfection possible by the combination with membrane filtration (UF)	• PAC must be disposed • Post-treatment required (membrane, textile or sand filter) to prevent discharge of PAC • Production of PAC needs high energy • Adsorption capacity may fluctuate with each batch
Granular activated carbon (GAC)	• High pharmaceutical removal • Full scale evidence on practicability • Additional dissolved organic carbon removal • No formation • An existing sand filtration can relatively easily be replaced by GAC • GAC can be regenerated	• Production of GAC needs high energy • Still under investigation if more activated carbon is needed compared to PAC • Less flexible in operation than PAC and ozonation to react to changes in wastewater composition • Adsorption capacity may fluctuate with each batch
Nanofiltration (NF) and Reverse Osmosis (RO)	• High pharmaceutical removal • RO can reduce salinity • Effective disinfection • Full rejection of particles and particle-bound substances	• High energy requirements • High investment and re-investment costs • Disposal of concentrated waste stream • Need for pre-treatment to remove solids

Source: (Rizzo et al., 2019[48]).

Figure 3.1 provides a rough estimation of the relative effectiveness and cost of different advanced wastewater treatment methods, based on a four-year study in Stockholm on the presence and effects of pharmaceuticals in the aquatic environment, and preventive measures and possible treatment methods (Wahlberg, Björlenius and Paxéus, 2010[50]). From Figure 3.1, the most economical and effective advanced treatment options are ozonation and activated carbon adsorption. This fining aligns well with published literature but It is worth noting that there is a lack of standardisation in the tests being used to determine removal efficiencies, which complicates the comparison of different studies.

Figure 3.1. Relative reduction efficiency to remove pharmaceuticals and cost comparison between different advanced wastewater treatment methods

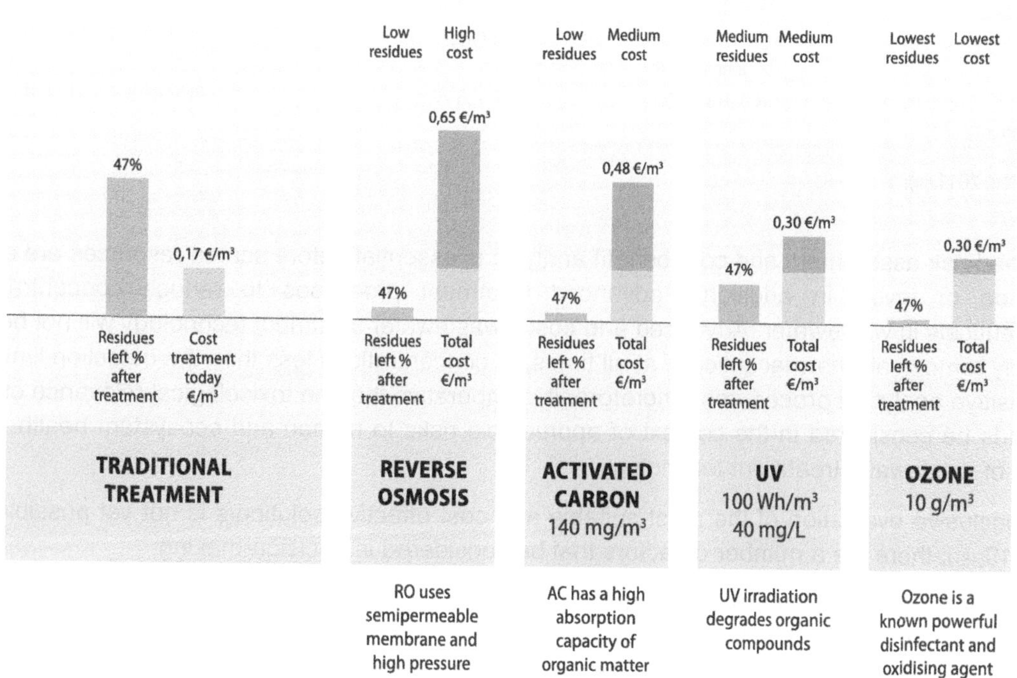

Note: Results presented are based on a four-year study in Stockholm on possible treatment methods to mitigate pharmaceuticals in wastewater.
Source: (Bui et al., 2016[51]), adapted from (Wahlberg, Björlenius and Paxéus, 2010[50]).

For removal of pharmaceuticals and complex mixtures, one single optimal wastewater treatment technology does not exist; several technologies are usually combined to try and cover the chemical profiles of different substances (Undeman and McLachlan, 2011[52]). Combinations of methods are often based on ozone and adsorption treatment (Cimbritz et al., 2016[53]; Bui et al., 2016[51]); adsorption processes (such as activated carbon) typically result in higher removal rates of hydrophobic and biodegradable chemicals, and oxidation processes (including ozone) typically result in high removal rates of reactive chemicals (Van Wezel et al., 2017[4]). Ozonation and activated carbon adsorption have a longer history as advanced technologies to upgrade traditional WWTPs in relation to other technologies (Bui et al., 2016[51]). In Switzerland, it was decided to implement advanced wastewater treatment at a large scale using ozonation and granulated activated carbon technologies.

Besides the efficiency (and cost) of removal of APIs and metabolites from wastewater, other environmental factors to consider in the selection of advanced wastewater treatment technologies include: potential formation of toxic transformation products, energy efficiency and carbon emissions, and disposal of any concentrate or sludge containing pharmaceutical residues (Bui et al., 2016[51]). Entec (2011[54]) provide an indication of these costs for a WWTP upgrade with GAC servicing 200,000 p.e. Other technical considerations include flexibility to deal with a range of pollutants and influent flows (accounting for demographic changes and increased rainfall variability with climate change), reliability and the quality of effluent, and ease of operation and maintenance requirements (Bui et al., 2016[51]).

Table 3.5. Estimated costs to the environment from WWTP upgrades

Environment impact	Estimated unit cost	Impact based on WWTP upgrade with GAC for p.e. 200,000	Total estimated annual cost per WWTP servicing 200,000 p.e.
Carbon emissions	EUR 59.8 / tonne	2,277 tonnes / yr	EUR 136,164 / yr
Additional sludge production	EUR 126-411 / tonne	394 tonnes / yr	EUR 49,644 – 161,934 / yr
Damage costs associated with additional energy use	EUR 0.018-0.059 / kWh	5,297,136 kWh / yr	EUR 95,348 – 312,531 / yr

Source: (Entec, 2011[54]).

An informed risk assessment and cost-benefit analysis is essential before scarce resources are allocated to upgrade or invest in additional advanced treatment processes to reduce concentrations of pharmaceuticals in wastewater. Advanced and costly wastewater treatment technology will not be able to completely remove all pharmaceuticals, at all times, to concentrations less than the detection limits of the most sensitive analytical procedures. Therefore, it is imperative that the toxicological relevance of various compounds be considered in the context of appreciable risks to human and ecosystem health, and the selection of wastewater treatment technology.

Whilst conclusive evaluation of the most suitable and cost effective solution/s is not yet possible (Rizzo et al., 2019[48]), there are a number of factors that be considered in decision-making:

- The allocation of advanced treatment techniques at WWTPs can be optimised with the aim to protect susceptible functions of the water system (such as protected areas or the provision of drinking water), or the upgrade of existing WWTPs as they reach the end of their life. For example, the Netherlands is performing a hotspot analysis to evaluate which WWTPs might deserve an extra purification step from the viewpoint of aquatic ecology and protection of drinking water. It is expected that this will be limited to a relatively small number of WWTPs.
- Opportunities to centralise treatment and close smaller, marginal WWTPs which are less cost-effective to upgrade (Table 3.6). Box 3.7 further illustrates the importance of economy of scale (centralisation of WWTPs) for improved cost-effectiveness, even for pollution hotspots such as hospitals; evidence shows that decentralised wastewater treatment at hospitals would not have a substantial impact on pharmaceutical loads entering centralised WWTPs, and finally the environment (Le Corre et al., 2012[55]).
- Site-specific limitations (e.g., availability of space and solar energy, cost of electricity) may lead to different conclusions for two different sites.
- Opportunities for co-benefits, such as the removal of other wastewater contaminants and expanding the possibilities of wastewater reuse. With these co-benefits may come additional market demands and revenue streams for infrastructure investment.
- Different relevant endpoints for a safe effluent discharge or reuse, such as CECs abatement, effluent toxicity, bacteria inactivation, by-products minimisation or abatement, antibiotic resistance control and treatment cost (Völker et al., 2019[49]) (Rizzo et al., 2019[48]).

Table 3.6. Cost comparison for removal of pharmaceuticals at different scales in Germany, Switzerland and the Netherlands

Euros per cubic metre

	Germany	Switzerland	Netherlands	Sweden
Small-scale WWTPs	€0.21 ± 0.08	€0.15–0.30	€0.22–€ 0.26 ± 0.05	
Medium-scale WWTPs	€0.19 ± 0.08		€0.18–€ 0.20 ± 0.05	€0.15–0.30
Large-scale WWTPs	€0.14 ± 0.08	€0.05–0.11	€0.16–€ 0.18 ± 0.05	

Note: Cost estimates include the following advanced wastewater treatment technologies: ozone + sand filtration, powered activated carbon + sand filtration, and activated carbon filtration.
Source: (Bui et al., 2016[51]).

Box 3.7. Decentralised advanced wastewater treatment at hospitals is less cost-effective than centralised municipal WWTP upgrades

Despite, the fraction of pharmaceuticals in wastewater discharged from hospitals being relatively low (15-20%) in comparison to the total load of municipalities, hospitals are a pollution hotspot for specialised pharmaceuticals with high environmental risks (e.g. X-ray contrast media, cytostatics, cancer treatments and some antibiotics) (PILLS, 2012[56]; Le Corre et al., 2012[55]). This offers the opportunity to eliminate high amounts of these specific pharmaceuticals from the environment with decentralised hospital wastewater treatment plants (PILLS, 2012[56]).

The early separation and on-site treatment of CECs (including pharmaceuticals) in wastewater from hospitals was investigated in Winterthur, Switzerland (Eawag, 2007[57]). While the study concluded that decentralised measures were capable of lowering emissions of CECs to the environment, it was also concluded that these approaches are not yet sufficiently developed to be cost-competitive with centralised end-of-the-pipe approaches (i.e. large-scale WWTP upgrades). The annual operation costs of a dedicated wastewater treatment with ozonation at the hospital would be as high as 40% of the annual operating costs of an additional ozonation stage at the central WWTP, while decreasing the overall CEC mass flow in the domestic effluent by only 20% (Eawag, 2007[57]).

The results of this study support more recent studies (e.g. (Le Corre et al., 2012[55])) which suggest that the risks of human exposure to the pharmaceuticals exclusively administered in hospitals are limited, and decentralised wastewater treatment at hospitals would not have a substantial impact on pharmaceutical loads entering centralised WWTPs, and finally the environment.

Other alternative methods to reducing the release of specialised pharmaceuticals used in hospitals to the water cycle include hospital protocols to safely dispose of pharmaceutical waste. For example, a pilot project in a Dutch hospital showed that patients are often willing to use disposable urine-bags during the time it takes for X-ray contrast media to leave their body (typically one day). Contrast media is often used in highly concentrated dosages, and they are inert and mobile which makes them hard to remove from wastewater (PILLS, 2012[56]).

Sources: (PILLS, 2012[56]; Le Corre et al., 2012[55]; Le Corre et al., 2012[55]).

The Swedish Government has recognised the adverse effects of certain pharmaceuticals on the aquatic environment and has mandated an evaluation of the cost-effectiveness of upgrading WWTPs. A summary of the evaluation results is presented in Box 3.8.

Box 3.8. The cost-effectiveness of end-of pipe removal of CECs, Sweden

In 2013, Swedish Parliament introduced a bill mandating the evaluation of advanced wastewater treatment technologies for the removal of pharmaceutical residues and other CECs by 2018. Because future impacts on the environment and human health are difficult to predict, the introduction of advanced wastewater treatment can be justified on the basis of the precautionary principle, as per the general rules in the Swedish Environmental Code.

The advanced wastewater treatment technologies examined included different treatment processes focusing on those sufficiently accessible and realistic to implement today, including ultra-filtration, ozonation, biologically active filtration, adsorption (pulverised activated carbon and granulated activated carbon) or combinations thereof.

The main results of the evaluation are as follows:

- A combination of different advanced wastewater treatment technologies result in >80% removal of pharmaceuticals. However, installation and operation of combined technologies result in higher costs compared to individual ones. Individual treatments can be justified depending on site-specific conditions or budget.

- Ozonation is the least expensive additional treatment step; however, in contrast to the Stockholm study by (Wahlberg, Björlenius and Paxéus, 2010[50]) (Figure 3.1), ozonation showed lower removal compared to biological and adsorptive methods. Adsorptive and biological technologies were the most energy-efficient technologies, with approximately 2-10% increased consumption. Technologies using oxidative treatments (e.g. ozonation) increase energy consumption by 20-60% and may also give rise to formation of residues with potential ecotoxicological effects.

- The costs of advanced treatment vary widely regarding the different capacities and size of WWTPs. Economies of scale and cost effectiveness can be achieved for advanced treatment upgrades at larger WWTPs. In general, larger facilities have more resources to ensure follow-up, process optimisation, and operation and maintenance of the facility.

- The cost of upgrading large WWTPs (>100,000 population equivalents) is estimated to be approximately 1 SEK/m^3. Upgrades of smaller facilities (2,000- 20,000 PE) can reach up to 5 SEK/m^3.

- The cost for upgrading each of the medium-large 431 WWTPs in Sweden (>2,000 PE) was estimated at between SEK 241 million to 2.1 billion per year, corresponding to SEK 55-480 per household per year. The cost estimates do not include costs associated with policy instruments, such as taxes, subsidies, fees, discharge allowances, permits and regulations.

The Swedish EPA identified five main factors for prioritising WWTP upgrades, including:

- The amount of pharmaceutical residues and other persistent pollutants released into receiving waters.

- The presence of several WWTPs that discharge to the same receiving water body.

- The water recharge rate of the receiving waters, where the receiving waters with low initial dilution and a low water recharge rate risk reaching the threshold values stated in the assessment criteria for river-basin-specific pollutants and effect levels.

- Fluctuations in water recharge rate over the year in the receiving waters, and variations in effluent volumes from the WWTP.

- The receiving water body's sensitivity, such as ecological sensitivity.

Source: (Swedish EPA, 2017[58]).

There are a number of considerations to be made when prioritising WWTPs for upgrade to advanced treatment that have the ability to maximise benefits and lower costs (including capital, operation and maintenance, opportunity and environmental costs). Firstly, it is important to invest in wastewater connection and treatment services where access is lacking, or treatment is limited (e.g. primary or secondary treatment only, and combined sewer overflow systems). Secondly, prioritisation should primarily be based on the concentration and volume of pharmaceutical residues entering water bodies, and their effects on ecological and human health. In relation to this, there are number of factors to consider for prioritisation of upgrades:

- WWTPs servicing a large population, or where urban population is projected to expand rapidly in the near future
- Age and state of WWTPs (i.e. WWTPs that are due for a refurbishment)
- Potential for amalgamating existing small WWTPs to one large centralised WWTP to achieve economy of scale
- WWTPs operating in basins where pharmaceutical manufacturing occurs
- Basins where several WWTPs and other sources of pharmaceutical pollution (e.g. pollution hotspots: aquaculture, intensive livestock agriculture, pharmaceutical manufacturing, hospitals) discharge to the same receiving water body.
- WWTPs upstream of drinking water supply sources
- WWTPs operating in basins where certain industries operate that require very high quality water sources, such as pharmaceutical and electronics industries
- WWTPs operating in basins under high water stress where wastewater reuse is a valuable resource (e.g. for irrigation on arable land)
- WWTPs operating in basins where the water recharge rate and dilution potential of the receiving waters is low
- WWTPs discharging to ecologically sensitivity water bodies
- WWTPs which discharge other persistent pollutants which may be effectively removed with advanced wastewater treatment (e.g. co-benefits of removing dissolved organic carbon, nutrient and pesticides, which are very costly to remove from raw drinking water supplies).

Modelling the impact of WWTPs on susceptible functions of water bodies can assist with decision making and allow for spatially smart implementation of WWTP upgrades. At the same time, trade-offs and associated costs should be considered and managed when deciding to upgrade WWTPs, including the aforementioned trade-offs such as: incomplete removal of pharmaceutical residues to varying degrees, depending on the treatment technology selected; the generation of toxic transformation products and sludge; increased use of energy and chemicals for treatment; increased carbon emissions; and increased capital and operating and maintenance costs, and potential affordability issues of sanitation bills.

Financing WWTP upgrades

The cost of WWTP upgrades, and the ability to finance them, is a concern for wastewater utilities. Table 3.7 provides a summary of the cost estimates of WWTP upgrades found in literature. Whilst these costs are high, it is helpful to understand how they compare to the cost of inaction, including the loss of ecosystem services. Section 1.4, chapter 1 highlights some of the impacts and costs of pharmaceuticals in the environment.

Table 3.7. Summary of the cost estimates for upgrading WWTPs to remove CECs

Region / country and source	Cost estimates and description
EU (Civity, 2018[59])	Cost of fourth treatment stage to remove pharmaceuticals: EUR 6.5 billion per year.
Finland (FIWA, 2016[60])	Pharmaceuticals and hazardous substances removal costs: EUR 700-1,400 million, with increase in operational by EUR 100–220 million. This corresponds to an increase of 20–30% of current wastewater fees. In addition, energy needs would increase by 30–80% from the current level
Germany (Entec, 2011[54])	Cost of removal of diclofenac: EUR 5-11 million per WWTP, with added energy and CO_2 costs. Additional sludge disposal costs totalling EUR 86 million – 256 billion per year. Average NPV capital and operating costs are EUR 398 per p.e. and 295 per p.e. for diclofenac removal using either GAC (99% removal) or UV (57% removal). Case study for WWTP in Ulm, with 440,000 population, cost: EUR 40 million.
Germany (Civity, 2018[59])	Cost of fourth treatment stage to remove pharmaceuticals: EUR 1.2 billion per year, or EUR 15.20 per person per year. Represents an average wastewater fee increase of 14-17%. Total cost over 30 years: EUR 36 billion.
Switzerland (Entec, 2011[54])	Diclofenac removal costs in 756 WWTPs: EUR 495 – 591 million capital costs, plus EUR 56-76m operational costs. Total costs EUR 0.03-0.3/year per m³ treated. Cost per person ranges from EUR 14,000 – 590,000.
Switzerland (Logar et al., 2014[61])	Estimated annual cost of upgrading 123 WWTP to remove CECs: CHF 133 million or CHF 86 per household connected. Discounted total costs over the 33 years life span of the investment: CHF 3.3 billion.
UK (Comber et al., 2007[62])	Cost of removal of pharmaceuticals for UK as a whole: EUR 10 billion, or potentially EUR 1.5 million per WWTP on average. This assumes all 6800 WWTPs need to be upgraded. Added costs for pharmaceutical removal per WWTP was EUR 0.8 – 25 million capital and EUR 0.02 – 4.1 million operational costs per year.
UK (Haigh, 2018[63])	Cost estimate to remove diclofenac, E1, E2 and EE2: GBP 27-31 billion over 20 years.
UK (Entec, 2011[54])	Cost of upgrading 48 WWTPs in Yorkshire Region: EUR 725 million, with operational costs of EUR 45 million per year. Discounted costs are EUR 1,020 million, discounted over 25 years. Scaling these figures up to England and Wales: EUR 12 – 14 billion.

Notes: WWTP: wastewater treatment plant; NPV: net present value; p.e.: population equivalent.
Source: Adapted from Ashley et al. (2018[64]), Financing water in 28 European countries: Challenges and responses. Background paper drafted for the OECD.

There are several options to finance WWTP upgrades, including: public taxes, wastewater tariffs, charges passed onto the pharmaceutical industry, or a combination thereof. Switzerland - the first country to tackle the CECs challenge at the national level – has introduced a wastewater tax to partially (75%) fund the upgrade of approximately 120 WWTPs to remove 80% of CECs (including pharmaceuticals) from wastewater by 2040 (Box 3.9). The drivers mentioned for the decision to upgrade WWTPs nationwide were: endocrine effects on the aquatic environment, risk of contamination of drinking water, the removal efficiency of new technology, societal acceptance and cost-efficiency of attributing advanced treatment techniques to upgrade WWTPs - although uncertainty remains (Eggen et al., 2014[65]; Logar et al., 2014[61]; Stamm et al., 2015[66]).

Box 3.9. Tax to update wastewater treatment plants to treat CECs, Switzerland

Switzerland has committed to remove 80% of CECs from wastewater by 2040. The Swiss Waters Protection Act requires polluted wastewater produced by households, businesses or industry to be treated before being discharged into water bodies. In 2014, the Waters Protection Act was revised, following agreement by Parliament, to further improve wastewater treatment for the removal of CECs (including pharmaceuticals). The revised Act involved three policy instruments: i) a new technical wastewater treatment standard, and ii) a nationwide wastewater tax, and iii) public subsidies to fund technical upgrades of WWTPs. The technical standard requires selected WWTPs to remove 80% of CECs from raw sewage, measured on the basis of 12 indicator substances, by 2040. The standard applies only to WWTPs that meet one of the following three selection criteria, which aim to achieve environmental improvements at acceptable costs:

- Large WWTP servicing > 80,000 population equivalents (p.e.);
- Medium-size WWTP (24,000-80,000 p.e.) that discharge into small rivers with low dilution ratio; and/or
- Medium-size WWTP (24,000-80,000 p.e) that discharge into water bodies used for drinking-water purposes.

In total, approximately 120 out of 700 WWTPs met one of the above three criteria for upgrade. It is projected that this will result in a 50% overall load reduction of CECs in surface water. In addition, several WWTPs will be closed and wastewater transferred to larger facilities where the treatment is considered to be more cost effective.

Pilot- and full-scale facilities assessed the effectiveness of various advanced wastewater treatment technologies, including ozonation, powered activated carbon, granular activated carbon, high-pressure membranes and advanced oxidation processes. Ozonation and powdered activated carbon showed the best applicability for Switzerland with the two techniques combined capable of removing 80% of detected CECs in wastewater.

The total investment cost to equip 100 WWTPs with advanced treatment technology was estimated to be CHF 1.2 b. Operation and maintenance costs were estimated to be an additional CHF 130 m per year, equivalent to 6% of the total current cost of wastewater treatment in Switzerland annually. The majority of the costs (75%) are financed by a new nationwide wastewater tax of CHF 9 per person per year, which is earmarked in a federal fund to upgrade WWTPs. The remaining 25% of costs are covered by the municipalities. As WWTPs are upgraded and become operational, the municipalities are exempted from the tax.

Despite having higher estimated costs than preventative source-directed measures, the end-of-pipe approach was selected because it was more predictable, measurable and feasible, and received support from industry, business, farmers, the research community and international actors. Furthermore, a national online survey indicated that the public were willing to pay the tax for reducing the potential environmental risk of pharmaceuticals; the average willingness to pay per household was CHF 100 per year, generating a total annual economic value of CHF 155 m per year.

Source: Summary of case study provided by Florian Thevenon, WaterLex International Secretariat, Switzerland. Additional source: (Logar et al., 2014[61]).

In Germany, the costs of nationwide WWTP upgrades to remove pharmaceuticals are estimated to amount to approximately EUR 1.2 billion per year, and in Europe, about EUR 6.5 billion per year (Civity, 2018[59]). In an investigation by Civity Consultants (2018[59]), preventive (source-directed and use-orientated) measures were deemed more cost-effective. However, if WWTP upgrades were to be considered, the researchers proposed that pharmaceutical companies should bear some of the cost and thus indirectly be incentivised to reduce entry of pharmaceutical residues to the wastewater system (Box 3.10).

Box 3.10. Proposed tax on pharmaceutical products as a financing mechanism for WWTP upgrades, Germany

The question of how to finance the installation of advanced treatment at selected WWTPs in Germany needs to be resolved, unless capital, operation and maintenance costs are to be paid for by wastewater tariffs and/or public taxes. A recent report by the Helmholtz Centre for Environmental Research (UFZ) advocates for the introduction of a regulatory tax on pharmaceutical products as an effective financing mechanism to upgrade WWTPs and to pass on some responsibility for the cost to industry.

A three-tier tax on pharmaceutical products was proposed:

1. High tax if it is clear that a pharmaceutical product causes damage to freshwater ecosystems.
2. Low tax if it is unclear whether a pharmaceutical product causes damage to freshwater ecosystems. Either the manufacturer or its delivery point (e.g. pharmacy) must pay a charge for the product's potential effect on water, as a precaution.
3. Zero tax if it can be proven that the pharmaceutical product does not result in any adverse environmental effects.

In theory, in order to qualify for zero tax, pharmaceutical companies would have to prove that their products were not harmful to the environment. Strict legal requirements would apply to assess the testing processes used by the manufacturers.

From a legal point of view, introducing such a regulatory tax on pharmaceutical products in Germany would not be an issue and may make good economic sense; tax revenue raised could provide adequate finance for WWTP upgrades, and the tax would act as an incentive for industry to develop green pharmaceuticals.

Source: (Gawel et al., 2017[67])

3.5.2. Pharmaceutical waste management and collection programmes

It is estimated that 10-50% of prescription medications are not taken as per the doctors' orders and unused or expired medicinal waste may be disposed of via the toilet - therefore offering zero therapeutic benefit and resulting in water pollution. Although the contribution of improper disposal of pharmaceuticals to the overall environmental burden is generally believed to be minor (Daughton and Ruhoy, 2009[68]), pharmaceutical collection schemes are still considered to be important.

Various systems have been developed around the world to recover and manage waste pharmaceuticals from households. Drug take-back programmes provide the public with a convenient way to safely dispose of leftover medications. In Europe, collection schemes of unused/expired medication are an obligatory post-pharmacy stewardship approach that reduces the discharge of pharmaceuticals into environmental waters (via WWTPs) and minimises the amounts of pharmaceuticals entering landfill sites.

High levels of public awareness and education on the environmental consequences of the disposal of unused/expired drugs are key for the success of collection programmes. Increased awareness and a change in consumer behaviour regarding disposal practices can be a cost-effective measure to help reduce environmental exposure to pharmaceuticals. The German education campaign: "No pharmaceuticals down the toilet or sink!" is considered a cost-efficient and effective reduction measure (UBA, 2018[2]). In the EU, the annual cost of pharmaceutical collection schemes range from EUR 250,000 in Belgium, up to EUR 15 million in Denmark (Entec, 2011[54]). Pharmacy on-site receptacles are considered the most common collection system.

Public collection schemes of unused pharmaceuticals are established in several OECD countries either as voluntary schemes or mandated by legislation (Table 3.8). Collection programmes are funded either by the government (e.g. Sweden, Australia) or by Extended Producer Responsibility (EPR) in line with the Polluters-Pays principle (e.g. in Canada, Belgium, Spain and France) (Barnett-Itzhaki et al., 2016[69]). Through EPR legislation, pharmaceutical companies are required to collect and dispose of the unused pharmaceuticals their companies put on the market. The advantage of EPR systems is that it takes the burden off the government and requires industry to finance and manage the collection and safe disposal (usually through incineration) of unused drugs. Companies can internalise these costs in the price of pharmaceuticals and can, in theory, provide services more cost-efficiently (for more on EPR see (OECD, 2016[70])).

Table 3.8. Household pharmaceutical collection and disposal programmes, select OECD countries

Country	Programme coverage	Method	Funding
U.S	28 local EPR laws in the US; 5 at state-level, 23 at local government level	Either voluntary programs by firms or governments, or mandatory programs through EPR	Governmental, Industry
Canada	Several regional programs across the country. Four EPR programs regulated under different jurisdictions	Retail pharmacies commonly act as collection sites	Brand-owners and contributions are based on market share.
Australia	National programme	Mandatory, Retail pharmacies commonly act as collection sites	Federal government
France	National programme	Mandatory EPR-scheme, Retail pharmacies commonly act as collection sites	Industry
Sweden	National programme	Mandatory EPR-scheme , Retail pharmacies commonly act as collection sites	Pharmacies

Source: Author

Key factors determining the success of public collection schemes include if the scheme is implemented by legislation (i.e. obligation to collect), and the level of public awareness (communication effectiveness). In Israel, there is no legislation regarding household medical waste collection and disposal; correspondingly, less than 14 % of Israelis return unused pharmaceuticals (Barnett-Itzhaki et al., 2016[69]). In Sweden, public collection schemes are mandatory and 75% of unused drugs are estimated to be returned (Larsson and Löf, 2015). Some Swedish pharmacies go as far as providing reward schemes and discounts when customers return unused drugs. One Swedish pharmacy chain has an annual 4 week- campaign to increase awareness that they collect pharmaceuticals. During the period after the campaign, the number of customers who returned old medicines tripled (Apoteket AB, 2018[71]). In the U.S., one of the biggest barriers to implementing pharmaceutical collection schemes is the need to change federal law and Drug Enforcement Administration regulations to allow pharmacies (the most convenient collection location) to collect from residents. This can be a very long process.

Regulations can also be designed to manage pharmaceutical waste from health care facilities Box 3.11 provides an example of such regulation in the U.S.

Box 3.11. Regulations on the disposal of hazardous pharmaceutical waste in the health sector, U.S.

In the U.S., the Management Standards for Hazardous Waste Pharmaceuticals (US EPA, 2019[72]) establishes a new rule for the healthcare sector (i.e. hospitals, veterinarian clinics, pharmacies) regarding the disposal of hazardous waste pharmaceuticals to protect human health and the environment. It aims to eliminate the intentional disposal of hazardous waste pharmaceuticals (both prescription and over-the-counter, non-credible and evaluated hazardous) to sewer systems. Hazardous waste pharmaceuticals must be disposed of in permitted hazardous waste facilities for combustion or incineration.

The new rule is in response to a growing body of studies documenting the presence of pharmaceuticals in drinking and surface waters, as well as their negative impacts to aquatic and riparian ecosystems. It is expected to reduce hazardous waste pharmaceuticals entering US waterways by up to 2,300 tonne per year.

A guidance document on managing hazardous pharmaceuticals waste, as well as best practices to minimise pharmaceutical waste, is provided for the healthcare sector (Practice Greenhealth, 2008[73]). It advocates, that "when in doubt, apply the Precautionary Principle"; when an activity raises threats of harm to human health or the environment, precautionary measures should be taken even if some cause and effect relationships are not fully established scientifically.

Sources: (Practice Greenhealth, 2008[73]) (US EPA, 2019[72]).

References

Anastas, P. and J. Warner (1998), *Green Chemistry : Theory and Practice*, Oxford University Press. [5]

Andrä, J. et al. (2018), "PharmCycle: a holistic approach to reduce the contamination of the aquatic environment with antibiotics by developing sustainable antibiotics, improving the environmental risk assessment of antibiotics, and reducing the discharges of antibiotics in the wastewater outlet", *Environmental Sciences Europe*, Vol. 30/1, p. 24, http://dx.doi.org/10.1186/s12302-018-0156-y. [14]

Apoteket AB (2018), *Lämna överblivna läkemedel till oss*, https://www.apoteket.se/om-lakemedel/overblivna-lakemedel/ (accessed on 11 September 2018). [71]

Ashley, R. et al. (2018), *Financing water in 28 European countries: Challenges and responses*, OECD Background Report, Paris. [64]

Barnett-Itzhaki, Z. et al. (2016), "Household medical waste disposal policy in Israel", *Israel Journal of Health Policy Research*, Vol. 5/1, http://dx.doi.org/10.1186/s13584-016-0108-1. [69]

Bartram, J. et al. (2009), *Water safety plan manual: step-by-step risk management for drinking-water suppliers*, World Health Organisation, Geneva. [16]

Bengtsson-Palme, J., L. Gunnarsson and D. Larsson (2018), "Can branding and price of pharmaceuticals guide informed choices towards improved pollution control during manufacturing?", *Journal of Cleaner Production*, Vol. 171, pp. 137-146, http://dx.doi.org/10.1016/J.JCLEPRO.2017.09.247.   [22]

BMEL (2019), *Press Release - Use of antibiotics in livestock farming is decreasing - Resistance situation improved (in German)*, https://www.bmel.de/SharedDocs/Pressemitteilungen/2019/135-Antibiotikaminimierungskonzept.html (accessed on 26 June 2019).   [44]

Boxall, A. (2012), *New and Emerging Water Pollutants arising from Agriculture*, OECD, Paris, https://www.oecd.org/tad/sustainable-agriculture/49848768.pdf.   [37]

British Poultry Council (2018), *Antibiotic stewardship: REPORT 2018*, British Poultry Council, http://file:///C:/Users/Sahlin_S/Downloads/BPC-2018-Antibiotic-Stewardship-Web.pdf.   [46]

Bui, X. et al. (2016), "Multicriteria assessment of advanced treatment technologies for micropollutants removal at large-scale applications", *Science of The Total Environment*, Vol. 563-564, pp. 1050-1067, http://dx.doi.org/10.1016/J.SCITOTENV.2016.04.191.   [51]

Cimbritz, M. et al. (2016), *Svenskt Vatten Utveckling Rening från läkemedelsrester och andra mikroföroreningar En kunskapssammanställning. Rapport Nr 2016-04*, Svenskt Vatten Utveckling, http://www.svensktvatten.se.   [53]

Civity (2018), *Costs of a fourth treatment stage in wastewater treatment plants and financing based on the polluter pays principle (in German)*, Civity Management Consultants, Berlin, https://www.bdew.de/media/documents/PI_20181022_Kosten-verursachungsgerechte-Finanzierung-4-Reinigungsstufe-_Klaeranlagen.pdf.   [59]

Comber, S. et al. (2007), *Dangerous Substances and Priority Hazardous Substances/Priority Substances under the Water Framework Directive*, UK Water Industry Research, London, https://ukwir.org/reports/07-WW-17-7/115386/Dangerous-Substances-and-Priority-Hazardous-SubstancesPriority-Substances-under-the-Water-Framework-Directive.   [62]

Daughton, C. (2014), "Eco-directed sustainable prescribing: Feasibility for reducing water contamination by drugs", *Science of the Total Environment*, Vol. 493, pp. 392-404, http://dx.doi.org/10.1016/j.scitotenv.2014.06.013.   [24]

Daughton, C. (2003), "Cradle-to-cradle stewardship of drugs for minimizing their environmental disposition while promoting human health. I. Rationale for and avenues toward a green pharmacy.", *Environmental health perspectives*, Vol. 111/5, pp. 757-74, http://dx.doi.org/10.1289/ehp.5947.   [8]

Daughton, C. and I. Ruhoy (2013), "Lower-dose prescribing: Minimising "side effects" of pharmaceuticals on society and the environment", *Science of The Total Environment*, Vol. 443, pp. 324-337, http://dx.doi.org/10.1016/j.scitotenv.2012.10.092.   [26]

Daughton, C. and I. Ruhoy (2009), "Environmental footprint of pharmaceuticals: the significance of factors beyond direct excretion to sewers", *Environmental Toxicology and Chemistry*, Vol. 28/12, p. 2495, http://dx.doi.org/10.1897/08-382.1.   [68]

Eawag (2007), *Eawag Annual Report 2007*, Swiss Federal Institute of Aquatic Science and Technology (Eawag), Dübendorf.   [57]

EC (2006), *Integrated Pollution Prevention and Control: Reference Document on Best Available Techniques for the Manufacture of Organic Fine Chemicals*, European Commission, Brussels, http://eippcb.jrc.es. (accessed on 1 February 2019). [19]

Eggen, R. et al. (2014), "Reducing the Discharge of Micropollutants in the Aquatic Environment: The Benefits of Upgrading Wastewater Treatment Plants", *Environmental Science & Technology*, Vol. 48/14, pp. 7683-7689, http://dx.doi.org/10.1021/es500907n. [65]

Entec (2011), *Technical Support for the Impact Assessment of the Review of Priority Substances under Directive 2000/60/EC*, Entec UK LImited, London, https://circabc.europa.eu/sd/a/d6673532-e862-4d40-ab5f-8aaffc766f58/DEHP.pdf. [54]

EU (2010), *Directive 2010/75/EU of the European Parliament and of the Council of 24 November 2010 on industrial emissions (integrated pollution prevention and control) (Recast)*. [18]

European Commission (2005), *Ban on antibiotics as growth promoters in animal feed enters into effect*, http://europa.eu/rapid/press-release_IP-05-1687_en.htm (accessed on 11 September 2018). [43]

FAO and Denmark Ministry of Environment and Food (2019), *Tackling antimicrobial use and resistance in pig production: Lessons learned from Denmark*, FAO, Rome. [47]

FIWA (2016), *Technical and economic review of wastewater treatment in Finland (in Finish)*, Finnish Water Utilities Association, Helsinki, http://www.vvy.fi. [60]

Gawel, E. et al. (2017), *Drug Levy - Compulsory Drug Surcharge for Measures to Reduce Micropollutants in Waters [In German]*, Umweltbundesamt [Federal Environment Agency], Dessau-Roßlau, https://www.umweltbundesamt.de/publikationen/arzneimittelabgabe-inpflichtnahme-des-arz (accessed on 1 February 2019). [67]

Government of Canada (2018), *Responsible use of Medically Important Antimicrobials in Animals*, https://www.canada.ca/en/public-health/services/antibiotic-antimicrobial-resistance/animals/actions/responsible-use-antimicrobials.html (accessed on 11 September 2018). [45]

Gustafsson, L. et al. (2011), "The 'Wise List': A Comprehensive Concept to Select, Communicate and Achieve Adherence to Recommendations of Essential Drugs in Ambulatory Care in Stockholm", *Basic & Clinical Pharmacology & Toxicology*, Vol. 108/4, pp. 224-233, http://dx.doi.org/10.1111/j.1742-7843.2011.00682.x. [30]

Haigh, N. (2018), *The Chemical Investigations Programme (CIP) and economic analysis to support policy solutions. Case study submitted to the OECD*, UK Department for Environment, Food and Rural Affairs (Defra). [63]

IACG (2018), *Antimicrobial resistance: Invest in innovation and research, and boost research and development and access. IACG discussion paper*, UN Interagency Coordination Group on Antimicrobial Resistance. [28]

Kemper, M., A. Hein and S. Lehmann (2017), *Veterinary medicine and the environment: How can veterinary medicine reduce entry to the environment? [In German]*, German Environment Agency (UBA), Berlin, http://www.uba.de/TAM-eintrag. [42]

Kümmerer, K. (2007), "Sustainable from the very beginning: rational design of molecules by life cycle engineering as an important approach for green pharmacy and green chemistry", *Green Chemistry*, Vol. 9/8, p. 899, http://dx.doi.org/10.1039/b618298b. [3]

Larsson, D. (2014), "Pollution from drug manufacturing: review and perspectives", *Philosophical transactions of the Royal Society of London. Series B, Biological sciences*, Vol. 369/1656, http://dx.doi.org/10.1098/rstb.2013.0571. [1]

Le Corre, K. et al. (2012), "Consumption-based approach for assessing the contribution of hospitals towards the load of pharmaceutical residues in municipal wastewater", *Environment International*, Vol. 45, pp. 99-111, http://dx.doi.org/10.1016/j.envint.2012.03.008. [55]

Leder, C., T. Rastogi and K. Kümmerer (2015), "Putting benign by design into practice-novel concepts for green and sustainable pharmacy: Designing green drug derivatives by non-targeted synthesis and screening for biodegradability", *Sustainable Chemistry and Pharmacy*, http://dx.doi.org/10.1016/j.scp.2015.07.001. [6]

Logar, I. et al. (2014), "Cost-Benefit Analysis of the Swiss National Policy on Reducing Micropollutants in Treated Wastewater", *Environmental Science & Technology*, Vol. 48/21, pp. 12500-12508, http://dx.doi.org/10.1021/es502338j. [61]

Marquardt, R. and S. Li (2018), "Antimicrobial resistance in livestock: advances and alternatives to antibiotics", *Animal Frontiers*, Vol. 8/2, pp. 30-37, http://dx.doi.org/10.1093/af/vfy001. [39]

Matus, K. et al. (2012), "Barriers to the Implementation of Green Chemistry in the United States", *Environmental Science & Technology*, Vol. 46/20, pp. 10892-10899, http://dx.doi.org/10.1021/es3021777. [10]

Muñoz, I. et al. (2009), "Chemical evaluation of contaminants in wastewater effluents and the environmental risk of reusing effluents in agriculture", *TrAC Trends in Analytical Chemistry*, Vol. 28/6, pp. 676-694, http://dx.doi.org/10.1016/J.TRAC.2009.03.007. [35]

noPILLS (2015), *Interreg IV B NWE project partnership 2012 - 2015 noPILLS report*, EU Interreg North-West Europe Programme, Lille, http://www.no-PILLS.eu. [27]

OECD (2018), *Best Available Techniques for Preventing and Controlling Industrial Pollution*, Environment, Health and Safety, OECD Environment Directorate. [17]

OECD (2018), *Stemming the Superbug Tide: Just A Few Dollars More*, OECD Health Policy Studies, OECD Publishing, Paris, https://dx.doi.org/10.1787/9789264307599-en. [29]

OECD (2016), *Extended Producer Responsibility: Updated Guidance for Efficient Waste Management*, http://dx.doi.org/10.1111/jiec.12022. [70]

OECD (2012), *The Role of Government Policy in Supporting the Adoption of Green/Sustainable Chemistry Innovations. OECD Series on Risk Management No. 26*, OECD, Paris, http://www.oecd.org/env/ehs/risk-management/reports-on-sustainable-chemistry.htm. [12]

PILLS (2012), *Pharmaceutical residues in the aquatic system: a challenge for the future. Insights and activities of the European cooperation project PILLS*, Emschergenossenschaft, Essen, http://www.pills-project.eu/PILLS_summary_english.pdf. [56]

Pope, L. et al. (2008), "Exposure assessment of veterinary medicines in terrestrial systems", in Crane, M., A. Boxall and K. Barrett (eds.), *Veterinary Medicines in the Environment*, CRC Press, http://dx.doi.org/10.1201/9781420084771. [36]

Practice Greenhealth (2008), *Managing Pharmaceutical Waste: A 10-Step Blueprint for Healthcare Facilities in the United States*, US EPA, https://www.epa.gov/hwgenerators/management-pharmaceutical-hazardous-waste (accessed on 26 June 2019). [73]

Rastogi, T., C. Leder and K. Kümmerer (2015), "Re-Designing of Existing Pharmaceuticals for Environmental Biodegradability: A Tiered Approach with β-Blocker Propranolol as an Example", *Environmental Science and Technology*, http://dx.doi.org/10.1021/acs.est.5b03051. [7]

Ringbom, T. (2017), *Tonnes of diclofenac in our water - rule change is needed (in Swedish)*, Läkartidningen, https://www.lakartidningen.se/Opinion/Debatt/2017/11/Sverige-slapper-ut-flera-ton-diklofenak-i-miljon/ (accessed on 26 June 2019). [33]

Rizzo, L. et al. (2019), "Consolidated vs new advanced treatment methods for the removal of contaminants of emerging concern from urban wastewater", *Science of The Total Environment*, Vol. 655, pp. 986-1008, http://dx.doi.org/10.1016/J.SCITOTENV.2018.11.265. [48]

SAICM (2015), *Nomination for new emerging policy issue: environmentally persistent pharmaceutical pollutants*, http://www.saicm.org/Portals/12/documents/meetings/ICCM4/doc/K1502367%20SAICM-ICCM4-7-e.pdf (accessed on 23 January 2019). [13]

Stamm, C. et al. (2015), "Micropollutant Removal from Wastewater: Facts and Decision-Making Despite Uncertainty", *Environmental Science & Technology*, Vol. 49/11, pp. 6374-6375, http://dx.doi.org/10.1021/acs.est.5b02242. [66]

Stockholm County Council (2014), *Environmentally Classified Pharmaceuticals 2014-2015*, Stockholm County Council, https://noharm-europe.org/documents/environmentally-classified-pharmaceuticals-2014-2015. [31]

Sweco (2016), *Report on the need for advanced treatment to separate pharmaceutical residues from wastewater, Sweden (in Swedish)*, Sweco Environment AB, Stockholm, https://www.naturvardsverket.se/upload/miljoarbete-i-samhallet/miljoarbete-i-sverige/regeringsuppdrag/2017/ru-rapport-behov-av-avancerad-rening-sweco.pdf (accessed on 26 June 2019). [34]

Swedish Association of Pharmacy (2018), *All pharmacies start to inform about diclofenac's environmental impact. Press Release 3 October 2018 [In Swedish]*, http://www.sverigesapoteksforening.se/alla-apotek-borjar-informera-om-diklofenaks-miljopaverkan/ (accessed on 5 February 2019). [32]

Swedish EPA (2017), *Advanced wastewater treatment for separation and removal of pharmaceutical residues and other hazardous substances Needs, technologies and impacts. REPORT 6803*, Swedish Environmental Protection Agency, Stockholm, https://www.naturvardsverket.se/Documents/publikationer6400/978-91-620-6803-5.pdf?pid=21820. [58]

Swedish Medical Products Agency (2018), *Proposal to reduce environmental impact from the manufacture of medicines and active pharmaceutical ingredients*, https://lakemedelsverket.se/overgripande/Om-Lakemedelsverket/Miljoarbete/Rapporter/ (accessed on 26 June 2019). [23]

Topp, E. et al. (2008), "Runoff of pharmaceuticals and personal care products following application of biosolids to an agricultural field", *Science of The Total Environment*, Vol. 396/1, pp. 52-59, http://dx.doi.org/10.1016/j.scitotenv.2008.02.011. [38]

U.S. Census Bureau (2008), *Pollution Abatement Costs and Expenditures 2005*, U.S. Government Printing Office, Washington, DC. [11]

UBA (2018), *Recommendations for reducing micropollutants in waters. Background- April 2018*, German Environment Agency, Dessau-Roßlau. [2]

UN Environment (2019), *Global Chemicals Outlook II: From legacies to innovative solutions*, United Nations Environment Programme, https://wedocs.unep.org/bitstream/handle/20.500.11822/27651/GCOII_synth.pdf?sequence=1&isAllowed=y (accessed on 27 June 2019). [9]

Undeman, E. and M. McLachlan (2011), "Assessing Model Uncertainty of Bioaccumulation Models by Combining Chemical Space Visualization with a Process-Based Diagnostic Approach", *Environmental Science & Technology*, Vol. 45/19, pp. 8429-8436, http://dx.doi.org/10.1021/es2020346. [52]

US EPA (2019), *Final Rule: Management Standards for Hazardous Waste Pharmaceuticals and Amendment to the P075 Listing for Nicotine, EPA-HQ-RCRA-2007-0932*, United States Environmental Protection Agency, Washington, DC, https://www.epa.gov/hwgenerators/final-rule-management-standards-hazardous-waste-pharmaceuticals-and-amendment-p075#rule-history (accessed on 26 June 2019). [72]

US EPA (2006), *Permit Guidance Document: Pharmaceutical Manufacturing Point Source Category (40 CFR Part 439)*, U.S. Environmental Protection Agency, Washington, https://www.epa.gov/sites/production/files/2015-10/documents/pharmaceutical-permit-guidance_2006.pdf (accessed on 1 February 2019). [20]

van Dijk, A. et al. (2018), "The potential for immunoglobulins and host defense peptides (HDPs) to reduce the use of antibiotics in animal production", *Veterinary Research*, Vol. 49/1, http://dx.doi.org/10.1186/s13567-018-0558-2. [40]

Van Wezel, A. et al. (2017), "Mitigation options for chemicals of emerging concern in surface waters operationalising solutions-focused risk assessment", *Environmental Science: Water Research and Technology*, Vol. 3/3, pp. 403-414, http://dx.doi.org/10.1039/c7ew00077d. [4]

Völker, J. et al. (2019), "Systematic Review of Toxicity Removal by Advanced Wastewater Treatment Technologies via Ozonation and Activated Carbon", *Environmental Science & Technology*, p. acs.est.9b00570, http://dx.doi.org/10.1021/acs.est.9b00570. [49]

Wahlberg, C., B. Björlenius and N. Paxéus (2010), *Pharmaceuticals: Presence and effects in the aquatic environment, preventive measures and possible treatment methods [In Swedish]*, Stockholm Water Company, Stockholm, http://www.stockholmvattenochavfall.se/globalassets/pdf1/rapporter/avlopp/avloppsrening/lakemedelsrapport_slutrapport.pdf. [50]

WHO (2015), *Global action plan on antimicrobial resistance*, World Health Organisation, https://www.who.int/antimicrobial-resistance/publications/global-action-plan/en/. [25]

WHO (2015), *Vaccinating salmon: How Norway avoids antibiotics in fish farming*, http://www.who.int/features/2015/antibiotics-norway/en/ (accessed on 11 September 2018). [41]

WHO (2014), *WHO good manufacturing practices for pharmaceutical products*, https://www.who.int/medicines/areas/quality_safety/quality_assurance/production/en/ (accessed on 26 June 2019). [21]

WHO (2012), *Pharmaceuticals in Drinking Water*, World Health Organisation, Geneva. [15]

4 Recommendations for the management of pharmaceuticals in freshwater

This final chapter presents the case for a life cycle, multi-sector approach to the management of pharmaceuticals in freshwater. Drawing from policy messages and case studies from the previous chapters, the report concludes with a set of recommendations on the management of pharmaceutical residues in freshwater for central government and other stakeholders.

4.1. Key messages: A life cycle, multi-sector approach to managing pharmaceutical residues in freshwater

No single policy instrument is capable of managing all sources of pharmaceutical pollution in freshwater. Likewise, there is no single culprit responsible for pharmaceutical pollutants reaching water bodies. In order to not only reduce existing known environmental threats, but to also minimise potential hazards, an approach for addressing pharmaceutical pollutants requires the involvement of central and local government agencies from various sectors (e.g. environment, agriculture, health, chemical safety), pharmaceutical industry, human and animal health care providers, patients and farmers, and water, wastewater and solid waste service providers. France, Germany, the Netherlands, Sweden and the UK have all established multi-stakeholder dialogues to address the pharmaceutical challenge.

An efficient abatement strategy combines policy options at various stages of the pharmaceutical life cycle, using source-directed, use-orientated and end-of-pipe measures. A focus on preventive options early in the pharmaceutical life cycle, may deliver the most long-term, cost-effective and large-scale benefits.

Three important actions to consider are: the promotion of green pharmacy and good manufacturing practices; the inclusion of environmental risks in the risk-benefit analysis of marketing authorisation for new pharmaceuticals; and post-authorisation monitoring and mitigation of high-risk pharmaceuticals (including of those already approved on the market). Data sharing and institutional coordination is necessary to reduce the knowledge gaps and increase efficiency at least cost to society.

4.2. A policy toolbox for a life cycle, multi-sector approach

The presence of pharmaceutical residues in water bodies is well documented and the hazards this creates are beginning to be understood (see Chapter 1). It is time to take effective action. Chapter 2 outlines the weaknesses of the current ERA process, and documents advances in monitoring and modelling to reduce uncertainties. Chapter 3 argues that current policy approaches in many countries to manage pharmaceutical residues in water are often reactive (i.e. measures are adopted only when risks can be proven and routine monitoring technologies exist), substance-by-substance (i.e. environmental quality norms for individual substances) and resource intensive. Such approaches are ill-adapted to emerging challenges, and the growing knowledge of environmental and health hazards triggered by pharmaceutical residues in freshwater ecosystems.

The large amount of substances, diverse entry-pathways into the aquatic environment and time-sensitive dynamics, underline the complexity of designing policies to manage pharmaceuticals in water. Solutions that are tailored for the reduction of point-source pollution (e.g. WWTP upgrades), for example, generally do not address the issue of diffuse pollution (typically from the use of veterinary pharmaceuticals in agriculture), and vice versa. Likewise, policies that deal with seasonal inputs do not necessarily solve the issue of constant discharges. Additionally, there exists a variety of goals across OECD countries, including the conservation of ecosystems, securing drinking water quality, or protecting recreational areas, which require diverse policy approaches.

The complexity of the issue suggests that there is no single-best policy instrument for this problem. Only a carefully designed package of policies has the potential to comprehensively reduce pharmaceuticals in water. This way, policies can be designed in proportion to the scale of the problem, collectively acting at different political levels and scales, and adopting different policy instruments in different sectors (Metz and Ingold, 2014[1]). Such an approach demands action throughout the pharmaceutical life cycle (source-directed, use-orientated and end-of-pipe measures) to minimise adverse effects on freshwater ecosystems and human health. Finally, voluntary measures alone will not deliver; economic and regulatory drivers are needed to incentivise action.

There are several mitigation options for water quality improvement in the pharmaceutical life cycle (Figure 4.1), including improvements in the design (e.g. green pharmacy), registration and authorisation, production, use and waste phases, and finally technological interventions of WWTPs (Van Wezel et al., 2017[2]). A focus on preventive options early in a pharmaceutical's life cycle, may deliver the most long-term and large-scale benefits (chapter 3 documents some effective preventative measures). Relying on end-of-pipe WWTP upgrades only is costly, energy intensive and toxic transformation products may be formed (Haddad, Baginska and Kümmerer, 2015[3]). However, in combination with source-directed and use-orientated approaches, extra treatment at the level of WWTPs may play a role in reducing human pharmaceuticals reaching the environment, particularly in light of growing demand for pharmaceuticals by society.

Figure 4.1. The pharmaceutical life cycle

Source: Author

Policies that cost-effectively manage pharmaceuticals for the protection of water quality and freshwater ecosystems rest on five strategies:

1. Reporting on the occurrence, fate, and risks of pharmaceutical residues in water bodies, consideration of environmental risks in the risk-benefit analysis pre-authorisation of new pharmaceuticals, and continued monitoring of high-risk pharmaceuticals post-authorisation (including of those already approved on the market).

2. Source-directed approaches to impose, incentivise or encourage measures in order to prevent the release of pharmaceuticals into water bodies;

3. Use-orientated approaches to impose, incentivise or encourage reductions in the inappropriate and excessive consumption of pharmaceuticals;

4. End-of-pipe measures – as a compliment to strategies 1-3 - that impose, incentivise or encourage improved waste and wastewater treatment to remove pharmaceutical residues after their use or release into the aquatic environment; and

5. A collaborative life cycle approach, combining the four strategies above and involving several policy sectors.

Regulatory, economic and voluntary policy instruments are all part of the toolkit that is needed to manage multiple sources of pollution from different stakeholders at different stages of the pharmaceutical life cycle. Stakeholder engagement through inclusive water governance and information is increasingly recognised as critical to secure support for reforms, raise awareness about water risks and costs, increase users' willingness to pay, and to handle conflicts. Policy-makers will need to factor in financing measures for the upgrade, operating and maintenance costs of wastewater treatment plants, as well as policy transaction costs to facilitate the transition from reactive to proactive control of pharmaceutical residues in water bodies.

A selection of key mitigation and policy options for different stakeholders at different stages of the pharmaceutical lifecycle is presented in Table 4.1.

Table 4.1. Selection of key mitigation options for different stakeholders across the life cycle of pharmaceuticals

Step in pharmaceutical lifecycle	Relevant stakeholders	Mitigation options
Cross-cutting	Government, Industry, Research organisations	Targeted monitoring and prioritisation of high-risk APIs Harness new innovations in water quality monitoring, modelling, scenario development and risk assessment Centralised database with regulatory oversight to share ERAs and environmental monitoring data of APIs Environmental quality norms / water quality standards
Design	Industry	Innovation in green pharmacy, biological therapies, personalised or precision medicines
Authorisation	Government, Industry	Legislation and standardised methodology for environmental risk assessment and incorporation into benefit-risk assessment of pharmaceutical marketing authorisation More stringent conditions for putting a pharmaceutical on the market that is of high-risk to the environment (e.g. increased risk intervention and mitigation options, eco-labelling, prescription only, post-approval monitoring)
Production	Industry, Government, Intergovernmental Organisations	Green public procurement with environmental criteria Environmental criteria for Good Manufacturing Practices, effluent discharge limits and disclosure of pharmaceutical wastewater discharge from supply chains[1]
Consumption (professional use)	Agriculture, Health sector, Government	Emission prevention through disease prevention and sustainable use of pharmaceuticals • improved human and animal health and well-being • improved diagnostics, avoided prescriptions • improved hygienic standards in health facilities, stable management and livestock handling • personalised medicines, vaccinations, targeted delivery mechanisms • prescription of environmentally-friendly pharmaceutical alternatives[2] • restrictions or bans of unnecessary high-risk pharmaceuticals (e.g. veterinary use of antibiotics for preventative measures and hormones as growth promoters in livestock)
Consumption (over-the-counter purchases/ self-prescription)	Health sector, Industry, Consumers	Eco-labelling of high-risk over-the-counter pharmaceutical products to improve consumer choice selection and awareness
Collection and disposal	Solid waste utilities, Industry	Education campaigns to avoid disposal of pharmaceuticals via sink or toilet Public pharmaceutical collection schemes for unused drugs Extended producers responsibility schemes Improved manure management by passive storage or anaerobic fermentation in biogas plants
Wastewater treatment	Wastewater utilities	Upgrade of wastewater treatment plants
Drinking water treatment	Drinking water utilities	Upgrade of drinking water treatment plants Water safety planning

1. GMP criteria may need to be redefined under the auspices of the WHO, including provision of globally harmonised environmental standards as part of the regulatory controls for pharmaceutical products. When negotiating environmental criteria for GMP, care would need to be taken to avoid withdrawal of countries from existing GMP agreements.
2. Requires that substitute pharmaceutical is available with lower environmental risk. An alternatives assessment would be required to confirm this, in order to prevent pollution-swapping.
Source: Author

4.3. The interlinkages between freshwater, pharmaceutical, and human and animal health policies

The life cycle approach refers to interlinkages between the control of pharmaceutical residues in freshwater and sector-specific policies to promote sustainable pharmaceutical industry and use. A summary of OECD recommendations on the management of pharmaceutical residues in freshwater is provided after the executive summary of this report. This section focuses on three sets of initiatives that can contribute to controlling pharmaceutical residues in freshwater at least cost to society:

- Improvements to environmental risk assessment and marketing authorisation of pharmaceuticals (section 4.3.1);

- Overcoming barriers to facilitate green pharmacy (section 4.3.2); and

- Opportunities to minimise costs related to data collection and bridging knowledge gaps (section 4.3.3). They can contribute to improving decision making under uncertainty, a feature of commonality of health-care and environmental policy (section 4.3.4).

Finally, equity should be considered in decision making regarding policies and investments to ensure that the needs of the most vulnerable populations, and that the allocation of costs, risks and benefits, are distributed in an equitable manner. This especially relates to pharmaceutical companies and manufacturers in two ways: i) a responsibility to prevent pollution and contribute to the costs of treating wastewater in line with the polluter pays principle, and ii) a responsibility to not simply outsource pollution to developing and emerging economies where environmental regulation and enforcement may be less stringent.

4.3.1. Improvements to environmental risk assessment and marketing authorisation of pharmaceuticals

One of the key steps from a regulatory view is to strengthen the ERA in the pharmaceutical marketing authorisation process. As mentioned in Chapter 2, current drug approval for human pharmaceuticals is based on safety, efficacy and quality; environmental effects are not considered in the risk-benefit analysis for marketing authorisation. This issue has been raised by several scientists (e.g. (Ågerstrand et al., 2015[4]) (Küster and Adler, 2014[5])) as well as within national action plans related to pharmaceuticals in the environment. The following improvements have been suggested:

- Improve the availability and transparency of ERA data and information. Establish a central database of ERA data with access rights to minimise duplication of testing (including animal testing) and improve consistency of ERAs. Such a database would require independent regulatory oversight.

- Use all available information subject to quality assurance and validation, and not only studies produced by the industry themselves.

- Develop a standardised methodology for integrated decision-making for assessing, comparing and communicating the therapeutic benefits and environmental risks for the benefit-risk assessment of pharmaceutical marketing authorisation.

- Assess environmental risk data requirements.

- Include environmental risks in the risk-benefit analysis.

- Include the risk potential of developing AMR in risk-benefit analysis.

- Ensure environmental risks are translated into enforced mitigation measures. Improve risk management options.

- Generate data and ERAs on existing pharmaceuticals on the market (i.e. pharmaceuticals authorised before ERAs became mandatory).

- Ensure environmental risks and impacts observed post-marketing are monitored and reported.

- Review ERAs on a regular basis to include new information, including data on actual use and effects, not just on default worst-case assumptions.

Considering the environment in benefit-risk analysis does not necessarily mean that pharmaceuticals should not authorised, rather that substances with high clinical importance, high environmental risk and no 'green' alternatives should be authorised with more strict monitoring (e.g. targeted monitoring, measured environmental concentrations, updated ERA with consumption data, spatial modelling) and environmental risk mitigation options (e.g. wastewater treatment at manufacturing plants, disposal instructions on packages, and prescription-only medicine). In rare cases when there are no appropriate measures available to minimise a serious environmental risk, and when other medicinal products or medical treatments are available on the market that offer adequate, equivalent health care, the marketing authorisation could be rejected based on environmental risk (UBA, 2018[6]). For more details on ERA and pharmaceuticals authorisation, refer to section 2.2.

4.3.2. Overcoming barriers to facilitate green pharmacy

A promising medium-term approach to reducing the environmental risks of pharmaceuticals, and possible health risks at source, is the rational design and manufacture of new green pharmaceuticals. Green pharmacy or 'benign by design' (Kümmerer, 2007[7]) is often referred to as: i) the development of new substances that are more efficiently biodegraded but retain their effective pharmaceutical properties, or ii) the re-design of existing pharmaceuticals for environmental biodegradability. The expected outcome is better biodegradable and pharmacologically active drug molecules that do not accumulate in, or cause adverse effects to, the environment.

Although research on green pharmacy has expanded in recent years, the share of green pharmaceuticals on the market is still low, and an agreed definition of green pharmacy has not been reached. Barriers delaying immediate progress, and policy options to overcome them, are presented in Table 4.2.

Table 4.2. Common barriers and policy solutions to facilitate green pharmacy

Barrier	Policy solution options
Technical barriers: uncertainties, and lack of cross-cutting expertise, definitions, metrics and transparency of data	• Develop a realistic and risk-based definition of 'green pharmacy' agreed between environmental chemists, clinical chemists, drug discovery scientists and other relevant stakeholders. • Facilitate knowledge creation and accessibility, through R&D spending and shared databases • Provide technical assistance for implementation to drive innovation. Demonstrate the feasibility of new green pharmaceuticals as new business opportunities for industry.
Economic and financial barriers: costs, lack of incentives, and markets	• Facilitate access to data (including testing) to avoid costly replication. Revise medicine regulations to allow cross-referencing of ERA information. • Demonstrate feasibility of new green pharmaceuticals which can present new business opportunities for industry • Create tax incentives and access to inexpensive capital to drive innovation. A return on public investments in new pharmaceuticals should be considered when assessing subsidies for the private sector in pharmaceutical development. • As part of combatting AMR, address economic models of new antibiotics, which currently link profit (sales) with volume (consumption). Options may include partnerships, grants and seed funding to stimulate innovation, or prizes and tax concessions aimed to reward at the end of the development process. • Develop and implement evidence-based technical guidelines on sustainable procurement of pharmaceuticals. Integrate environmental criteria into good manufacturing practices utilised by authorities to pre-qualify pharmaceuticals for procurement • Implement eco-labelling schemes of green pharmaceuticals, with or without a green premium • Consider granting additional intellectual property protection, and thus longer period of exclusivity, to offset the costs of innovation and reduce financial risks (provided exclusivity can be tied to ongoing environmental risk management during period of exclusivity)
Regulatory barriers: a regulatory focus on risk control, rather than risk prevention	• Ensure environmental risks are adequately taken into account in marketing authorisation of new pharmaceuticals, and translated into enforced mitigation action to incentivise investment in green pharmacy • Allow for an easy or fast-track marketing authorisation process for green pharmaceuticals based on the biodegradability of APIs and their metabolites and transformation products after use

Source: Author

4.3.3. Data sharing and institutional coordination to reduce the knowledge gaps

Reaching a rational assessment of the risks posed by pharmaceuticals as environmental pollutants needs to be done with a minimum investment of resources, which means avoiding reinvention and rediscovery of environmental testing and risk assessment. In order to address knowledge gaps and perform robust ERAs, it is necessary to harmonise data types and forms, and share existing information. Open source, good quality databases (with independent regulatory oversight), efforts to link databases to toxicity and exposure, and greater collaboration between stakeholders and across borders are called for in this regard (Ågerstrand et al., 2017[8]). In particular, collaboration between the traditionally separated environmental, chemical and medical sciences has a critical role to play. Collaborations among the environmental, chemical and medical sciences are important because in the final analysis, human health, animal health and the health of ecosystems are intimately tied, and in many respects, indistinguishable.

The following list of recommendations can facilitate harmonisation across political boundaries:

- *Define protocols used in data collection* for quantifying pharmaceutical usage by substance, and for determination of potential adverse effects on human and ecosystem health (including AMR and mixture effects).

- *Define common indicator substances*. With more precise and diverse monitoring approaches, an even greater amount of pharmaceuticals will be found in water bodies or living organisms, which again create new demands on research. At the same time it is a challenge for policymaking to deal with such volumes of scientific results and to adapt to new research developments. Indicator substances are single, defined substances which are easy to monitor due to existing knowledge on analytic methodologies. The detection of an indicator substance in a waterbody reveals the presence of more substances and, thus, points to a larger pollution issue. The advantage of an indicator approach is that costs and complexities are kept to a minimum, and a solid knowledge base is built for further pollution reduction measures.

- *Define uniform standards for data quality and storage*. Internationally coordinated policy guidelines, which clearly define data gathering and storage methods, and set baselines on what substances to scrutinise, is a necessary policy step in the process of building a robust knowledge base on pharmaceuticals, and modelling future scenarios of concentrations of pharmaceuticals in water.

- *Define common standards on how to prioritise contaminants and susceptible water bodies*. Because time and resources are not infinite, research must focus on the pharmaceuticals that represent the greatest threat, to the most sensitive and susceptible water bodies and ecosystems. The relative risk of pharmaceuticals should also be compared with other pollutants (e.g. heavy metals, persistent organic pollutants and other contaminants of emerging concern) to achieve improvements in water quality and ecosystems in the most cost effective way. Defining common standards and ranking approaches on how to prioritise contaminants and susceptible water bodies could be helpful in this regard. Several prioritisation approaches have been developed in academia to support decision-making (e.g. (Donnachie, Johnson and Sumpter, 2015[9]; Guo et al., 2016[10]; Roos et al., 2012[11]).

- *Establish data exchange platforms*. The exchange of, and access to, data of standardised quality is of utmost importance in order to improve knowledge on active pharmaceutical ingredients. It would be a fruitful contribution by the policy community to install a harmonised platform, or an international registry system, where data on pharmaceuticals and other emerging pollutants is stored and available for further research. Such a policy initiative would have to define how public authorities, research institutes, and private companies report data into the platform, and at which time intervals. Several efforts are underway to create platforms for sharing data[1].

- *Harmonise policy guidelines on analytical methods and risk assessment of pharmaceuticals*. Research on analytical methods and risk assessment regarding pharmaceuticals is a growing field; its geographical and disciplinary diversity is important for research innovation. However, it is necessary to integrate knowledge and gain an overview about the latest developments regarding monitoring methods from industry. To that end, existing repositories, data and published literature on pharmaceuticals are often underutilised. And in many cases, the data (ecotoxicity, properties and sales figures) for pharmaceuticals are confidential. In addition, ERAs of pharmaceuticals are not collected in any standardised way and are generally not accessible or searchable in a database. Harmonised analytical methods for ERAs and a coordinated higher-level synthesis of monitoring programmes and risk assessment would be useful for policy making. Synergies with existing harmonisation programmes can be used, such as the OECD's work on Test Guidelines or the Working Party on Hazard Assessment, RiBaTox[2], which assists the ERA, prioritisation and mitigation of contaminants. Other examples include the NORMAN[3] network and the EU SOLUTIONS project, which are both working on developing new, easier and more cost-efficient chemical analytical methodologies and software tools. The OECD Mutual Acceptance of Data (MAD) system[4] goes some way to achieving the exchange of information between OECD countries and produces savings from the reduction of duplicative testing for the assessment of new pharmaceuticals (OECD, 2019[12]).

4.3.4. Decision- and policy-making under uncertainty

Uncertainty is pervasive in health-care and environmental decision making (OECD, 2005[13]). A key challenge is to combine an evidence informed policy making approach with the need to make decisions under conditions of unpredictability, uncertainty and complexity. In health care and environmental management, the stakes for decisions are high, and may carry financial, health and environmental risks and rewards. How can decision making be transformed to cope with uncertainty and avoid paralysis? Evidence, including information on whether a new policy or technology presents value for money, plays a key part in aiding decision makers to make informed choices.

Opportunities to enhance understanding (and reduce uncertainties) on pharmaceuticals in the environment include: harmonisation of environmental monitoring and risk assessment approaches; better data quality and gathering; forecasting and scenario development; heightened transparency and sharing of information; integrated planning; and improved accessibility to tools and guidance. However, evidence is not always available to make informed decisions, and even when it is, some uncertainty will remain. The overarching imperative and responsibility for decision makers is to make decisions, even if on poor quality evidence. To defer consideration of a matter until the perfect evidence is available is, in effect, to decide.

As part of the ERA process and the authorisation of new pharmaceuticals, it is important to capture uncertainties and factor them into decision-making and development of policy responses to mitigate adverse environmental effects. Precautionary measures should be explored when scientific evidence is uncertain and when the possible consequences of not acting are high. For example, it is worthwhile considering a more proactive policy approach where future concerns are anticipated before any major environmental, health or economic consequences are felt, particularly because the damage caused at the population and ecosystem levels can take years to repair. This is particularly relevant to pharmaceuticals identified as high-risk as part of the marketing authorisation process (or post-authorisation) and the development of mitigation measures to minimise environmental impact.

Action should be in line with wider development objectives of safeguarding consumer, health and environmental protection, and supporting the principles of circular economy. Measures taken should be proportional, non-discriminatory, consistent with comparable measures, based on an examination of the potential benefits and costs of action or lack of action, subject to review, and capable of assigning responsibility for producing the missing scientific evidence. Scenario development can aid in decision-making, and exploring plausible benefits and consequences of precautionary measures and a range of policy options (Box 4.1).

> ## Box 4.1. Scenario development and testing as a means to manage uncertainties
>
> Despite the uncertainties surrounding pharmaceuticals in the environment (e.g. ecotoxicological effects, population dynamics, new pharmaceutical development, innovations in diagnostics and health care services, global environmental change), scenario development and resilience modelling can help in planning and testing long-term policy options for water security, and healthy humans, animals and environment (OECD, 2018[14]). Scenarios can inform today's thinking about strategic decisions through exploring different possible futures. This allows decisions to be stress-tested against whether they lock-in trajectories towards less desirable end states, and/or consideration of strategies that are robust to alternative directions (OECD, 2018[14]). Potential scenario choices may reflect:
>
> - improved attitudes toward sustainability, circular economy and green pharmacy
> - an innovative future in health services and diagnostics, reduction in chronic disease, and wastewater treatment technologies
> - various IPCC climate change scenarios which effect: temperature and rainfall variability; stress on ecosystems and the services they provide; frequency and intensity of natural disasters; disruption of wastewater, drinking water and health care services; and the occurrence, rate and spread of disease
> - financial constraints that for whatever reason may impede investment opportunities in WWTP and improved environmental performance
> - uncontrolled consumerism where economic growth, intensive agriculture and aquaculture production, and health care is pursued with little regard for the environment or social equity
> - various population scenarios affecting pharmaceutical consumption, including level of ageing, life expectancy and urbanisation
> - business as usual, extrapolating current trends and policy approaches.
>
> Scenario development needs to be an inclusive and participatory activity, to ensure that a wide range of stakeholder views are taken into account; this will make the scenarios more robust and realistic (OECD, 2018[14]).

Data and integrated planning with stakeholders at the basin scale can help set the ambition for improved water quality, and provide the basis for understanding the total financial spend requirements, and the relative priorities of different objectives. The full breadth of the evidence base can be aggregated so that a complete picture is presented of the state of the environment in the basin, together with current and future pressures and how these interact on society and the economy. Integrated planning can help to determine the type of policies to mitigate adverse effects, including the type and level of charges to recover costs, based on polluter or user pays principles to recover costs. Inevitably, this process will result in trade-offs, but these can be understood and arrived at through a negotiated process with stakeholders (OECD, 2018[14]).

Decision makers need not be limited to "all or nothing" approaches. There are opportunities for low regret options or staged roll-out of new policies. Options which are scalable and can be adopted incrementally will be able to respond better as more certainty emerges about the future. Low regret options may include: prevention of infection with improved sanitation and hygiene practice and education (therefore reducing the need for antibiotics and other pharmaceuticals); reduction of unnecessary use and release of antibiotics and hormones used for preventative measures and as growth promoters in agriculture and aquaculture; reduction of self-prescription and illegal sales of pharmaceuticals; and reduction of unknowns on relationships between pharmaceuticals, and human and environmental health. A summary of short- and medium-term measures for reduction of pharmaceuticals in the environment is provided in Table 4.3.

Table 4.3. Short- and medium-term options for reduction of pharmaceuticals in the environment

Short-term options	Long-term options
Develop monitoring programmes and incidence reporting of APIs in the environment, and their impact on human and ecosystem health. Utilise innovative monitoring and modelling techniques and methodologies	Review and revise the pharmaceutical authorisation process and ERA guidelines to improve environmental risk management of APIs. Include consideration of environmental risks in the risk-benefit analysis of authorisation of new pharmaceuticals. Ensure identified environmental risks are translated into enforced mitigation measures
Assess, map and prioritise APIs and water bodies of highest concern. Model projected risks based on future trends and potential scenarios in population dynamics	Require inclusion of ERA and effluent discharge data in sustainability and environmental reports of pharmaceutical manufacturing companies
Ensure identified environmental risks and impacts post-marketing are reported and ERAs are updated. Establish a centralised database with independent regulatory oversight to share ERA and API data	Factor pharmaceutical risks into drinking water safety plans
Avoid unnecessary prescriptions with improved diagnostics	Develop environmental criteria for green public procurement of pharmaceuticals
Avoid/ban/restrict unnecessary treatment (e.g. veterinary antibiotics for preventative measures and hormones as growth promoters)	Develop environmental criteria for good manufacturing processes. Reduce discharge of APIs in effluent from pharmaceutical manufacturing plants to below safe levels (e.g. PNEC concentrations)
Improve hygienic standards in hospitals and stables and other farming activities	Investigate eco-labelling of over-the-counter pharmaceutical products to improve consumer awareness of environmental impact
Roll out education campaigns to avoid disposal of pharmaceuticals via sink or toilet, or inappropriate animal manure management	Investigate the feasibility of prescription of environmentally-friendly pharmaceutical alternatives
Establish regular education and training of human and animal health practitioners and staff	
Improve health and well-being as a preventative measure (e.g. prophylactic vaccination)	
Incentivise green pharmacy, biological therapies, personalised and precision medicines. Research galenics to maximise absorption of drugs and minimise the excretion of APIs	
Mandate public take-back schemes for unused pharmaceuticals	
Develop technologies to remove pharmaceuticals in WWTPs, and cost-benefit analysis and financing options	

Notes: API: Active pharmaceutical ingredient. ERA: Environmental Risk Assessment. PNEC: Predicted no effect concentration. WWTP: Wastewater treatment plant.
Source: Author

4.4. A life cycle, multi-sector approach: Experience from selected OECD countries

Several countries have developed national action plans to address pharmaceuticals in the environment and have started a multi-sector dialogue to tackle the problem. At the EU level, a *Strategic Approach to Pharmaceuticals in the Environment* places an emphasis on sharing good practices, on cooperating at international level, and on improving understanding of the risks. It identifies actions for stakeholders throughout the pharmaceutical life cycle.

Common denominators in these plans are that actions and recommendations place a high importance on the exchange of data and knowledge between different sectors, and on education and communication. Each of these plans advocate for action throughout the life cycle of pharmaceuticals, with a strong emphasis on source-directed and use-oriented approaches (as opposed to end-of-pipe treatment options).

4.4.1. Germany: a multi-stakeholder dialogue to reduce contaminants of emerging concern in water

Germany has developed a multi-stakeholder dialogue to facilitate action on The Trace Substance Strategy. The dialogue was established 2014 on behalf of the Federal Health Ministry, coordinated by the Federal Environment Agency and with contributions from the Federal Institute for Drugs and Medical Devices. A key output of the dialogue was *Recommendations for reducing micropollutants in water* (UBA, 2018[6]), which provides specific recommendations to reduce human and veterinary pharmaceuticals in the environment. A summary of the recommendations is provided in Table 4.4. Some of these actions offer effective and immediately-feasible options to reduce human and veterinary pharmaceuticals in the environment, but most are expected to become effective only in the medium- to long- term horizon.

Table 4.4. Assessment matrix of selected source-directed and use-orientated measures to reduce human and veterinary pharmaceuticals, Germany

Measures	Effectiveness	Specific or Broad spectrum	Costs	Horizon	Feasibility
Developing and harmonising risk-reduction measures within the authorisation process	-/0	S	0/+	2	+
Banning PBT and vPvB substances in veterinary medicinal products	+	S	n.d.	2	-
Researching environmentally friendlier APIs or dosage forms	-	S	-	3	0
Target-group specific communication and information	0	B	+	2-3	+
Running information campaigns on the proper disposal of unused drugs	+	B	+	2-3	+
Over-the-counter regulatory system for APIs	+	S	0	3	+
Considering widening the requirements for prescription taking into account environmental concerns	n.d.	S	0	n.d.	0
Research on how modifying 'the right to dispense' many potentially affect the use of veterinary pharmaceuticals	n.d.	S	0	n.d.	0

Note: Expected effectiveness: (+ high), (0 moderate), (- low), (n.d. no data, uncertain), (S: measure is substance-specific), (B: measure has a broad spectrum effect), Costs: (+ low), (0 moderate), (- high), (n.d. no data, uncertain), Effectiveness horizon: (1 = short term < 5 years), (2 = medium term < 10 years), (3 = long term > 10 years), Feasibility: (+ immediately feasible), (0 not yet immediately feasible), (- still clear deficits/need for action (need for research, funding or acceptance).
Source: (UBA, 2018[6]).

4.4.2. France: Priority actions to reduce pharmaceuticals in water and empower local stakeholders[5]

In 2016, the second French National Plan against Micropollutants was launched as *The National Plan against Micropollutants 2016-2021* (Ministry of Ecological and Solidarity Transition, 2014[15]). The Plan builds on three objectives: i) reduce micropollutants, ii) acquire knowledge, and iii) prioritise action. Some of the key actions in the plan that specifically target pharmaceutical residues include:

- Implement the national guide on handling pharmaceutical waste and liquid waste in healthcare facilities
- Assess the management of unused pharmaceuticals in healthcare facilities and suggest evolutions
- Assess the relevance of the Swedish ranking of active substances (Box 3.2, chapter 3) based on their impact on the environment and the acceptability of such a ranking for pharmaceuticals by health professionals in France
- Assess mixture effects of micropollutants on aquatic flora and fauna, especially those linked with endocrine disruption
- Work on data sharing to improve knowledge of hazards and exposure regarding human and veterinarian pharmaceutical residues in waters
- Derive threshold values and methodologies to better assess water quality taking into account endocrine disruptors and relevant metabolites
- Identify metabolites of pharmaceutical products and assess analytical capacities in order to establish an early monitoring system.

Despite its non-binding nature, the Plan is a first step in the preparation of a broader toolbox of legally-binding policies in the future.

Recognising that CECs may not be great candidates for classic water quality regulation, the French Ministry of Ecological and Solidarity Transition established a five-year (2013-2018) subsidy programme (EUR 10 million) aimed at stimulating new innovative projects to manage CECs and empower local stakeholders.

A total of 13 projects were selected, targeting various stages of the life cycle of CECs, including management of: domestic point source pollution, health-related practices, and diffuse source pollution. A common denominator of the projects was that they all focus primarily on source-oriented solutions, a strategy that the Plan emphasises. All projects include solutions for better diagnostics, cost-efficient reduction of CECs and changes in practices of various stakeholders. In addition to encouraging innovation, the subsidy programme provides a platform promoting collaboration between various stakeholders in order to create an integrated strategy. It also takes into account socio-economic aspects to encourage stakeholders to accept practice changes. While the exercise has shown that there is potential for innovation at the local level, communication of the benefits and replication at the national scale remain a challenge.

4.4.3. Sweden: Increased consideration of the environmental risks of human pharmaceuticals

In Sweden, Parliament approved "Greater environmental considerations in international and EU pharmaceutical legislation by 2020" (now extended to 2030). Four specific measures are considered appropriate to reduce the environmental impact caused by the production and use of pharmaceuticals: including i) increase access to information on the impact of medicinal products; ii) establish more appropriate and better environmental testing, and revise the ERA guidelines; iii) consider environmental risks in risk-benefit authorisation of human pharmaceuticals in order to manage risk mitigation; and iv) establish mandatory minimum requirements for good pharmaceutical manufacturing practices.

In 2018, the Government commissioned the Swedish Medicine Agency to establish a knowledge centre for pharmaceuticals in the environment. The centre aims to gather Swedish actors and provide a platform for dialogue and cooperation. A budget of SEK 5 million has been allocated by the government (Government Offices of Sweden, 2017[16]).

4.4.4. United Kingdom: A "whole catchment" approach to managing pharmaceuticals in the environment[6]

In 2010, the UK Water Industry Research programme, with the support and collaboration of government and environmental regulators, initiated a multi-million pound Chemical Investigation Programme (CIP) into the scale of challenges to meeting existing Environmental Quality Standards (EQSs) detailed in the WFD, as well as emerging concerns such as pharmaceuticals.

The objectives of the CIP are to: i) gain definitive evidence of the true extent of discharges from WWTPs of both currently regulated chemicals and those of emerging concern; ii) explore mitigation options, such as new technologies; and iii) appraise options including their economic and environmental costs. In parallel, the UK Environment Agency and the Department for Environment, Food and Rural Affairs have been developing evidence on the economic impact of chemicals in treated effluents and receiving water bodies, and the benefits of mitigation. Fourteen pilot trials of new wastewater treatment processes are being conducted as part of the CIP (e.g. ozone, sand filtration, membrane bioreactor). The CIP is due for completion in 2020.

General preliminary findings of the CIP, which may have relevance to the future management of pharmaceuticals in the UK, include:

- A number of pharmaceuticals are statistically likely to be exceeding PNECs in the water environment[7]. Diclofenac, ibuprofen, EE2, propanol, erythromycin, azithromycin, clarithromycin and ranitidin were all detected at universal, or near-universal, high levels with high or very high severity, and with a high confidence of exceedance of PNEC. Other pharmaceuticals detected with a significant number of face value fails were E1, E2, ciprofloxacin, fluoxetine and metformin.

- The source of most pharmaceuticals is domestic. However, whole catchment studies have revealed that WWTP effluent is not always the main contributor. Water quality upstream of WWTPs can be poor due to diffuse sources of pollution. This suggests that catchment or source-directed approaches may be required for basic efficacy, rather than relying on end of pipe mitigation.

- Previous regulatory source-directed measures implemented in the UK have worked for other CECs. Notably, time series data demonstrates that concentrations of brominated diphenylethers (flame retardant) in WWTP effluents is declining by approximately 30% every five years in response to a ban of brominated diphenylethers, which was implemented over a decade ago. This clearly has implications for future investment in treatment technology; pharmaceuticals may also respond more quickly to source-directed approached than more ubiquitous legacy contaminants.

- The pilot trials of advanced wastewater treatment processes showed that removal of hormones was consistently good across all trialled processes, but removal of other pharmaceuticals was seen as variable. The new technologies currently remain subject to technical issues and contaminant removal mechanisms are not fully understood.

- Concerns remain within the water industry and regulatory agencies that advanced wastewater treatment processes are expensive to construct, operate, and maintain (in terms of both money and carbon). A high-level preliminary estimate of the costs of widespread "end of pipe" investment to tackle particular pharmaceutical concerns in the UK was made in 2013 at GDP 27-31bn over 20 years. Evaluating these costs with respect to the benefits is challenging, not least because of the underlying uncertainties surrounding the adverse effects to flora, fauna and humans of pharmaceuticals present in the water environment.

4.4.5. The Netherlands: "Chain approach" to pharmaceutical residues in water[8]

In the Netherlands, a holistic "chain approach" is being used to address the issue of pharmaceutical residues (both human and veterinarian) in water. The programme started in 2016 and considers the entire cycle, from the source to the end of the pipe, and supports various stakeholders in their voluntary efforts to reduce pharmaceutical pollution in water. When initiating the programme, four 'rules of the game' were established and agreed upon: 1) patients must keep access to the medicines they need (i.e. medicines shall not be banned), 2) all actions taken in the pharmaceutical chain should have a pragmatic approach and should be aimed at solving problems (measures for the sake of appearances to be avoided), 3) all stakeholders act where they can, within acceptable costs, and 4) stakeholders should not wait for other stakeholders to take the first step.

The two main drivers behind the programme were improved water quality and protection of drinking water. It is estimated that 140 tonnes of pharmaceuticals are discharged from WWTPs each year into Dutch waters. The programme links together the health care and water sectors[9] in the Netherlands. Although striving for the same goals, it quickly became clear that stakeholders were unfamiliar with each other's worlds, pinpointing the importance of cross-sector collaboration. An ongoing discussion is being held about the costs of potential measures and who should pay. This has raised questions regarding the applicability of the Polluter Pays Principle and who should be considered the polluter. Is it the patient who excretes the pharmaceutical residues, is it the doctor who prescribed the pharmaceuticals, the pharmacist that delivered them, or the industry that designs and produces them?

A total of 17 possible measures to reduce or mitigate the impacts of human pharmaceutical residues in water has been identified for further investigation (Table 4.5). The challenge will be to take measures at all relevant stages of the pharmaceutical chain, and to keep the enthusiasm that all stakeholders have shown to date.

Table 4.5. Examples of possible measures to reduce medicine residues at different stages of the pharmaceutical chain identified by the Netherlands

Possible measure	Stage in the pharmaceutical chain	Sector responsible
Identify pharmaceuticals that have negative environmental effects	Environmental monitoring	Water authorities and drinking water sector
Identify effects of veterinary pharmaceuticals in water	Environmental monitoring	Water authorities
Quantify emissions of veterinary pharmaceuticals to surface water and groundwater	Environmental monitoring	Several (new chain)
Develop 'green medicines' that have less environmental impact	Development & authorisation	Pharmaceutical companies and research institutions
Develop management system for environmental risks of medicines (Eco Pharmaco Stewardship)	Development & authorisation	Pharmaceutical companies
Improve access to environmental data on APIs	Development & authorisation	Pharmaceutical companies, authorising agencies, (international) authorities
Identify pairs of pharmaceuticals with same medic effect, but different environmental impact	Prescription & consumption	Several; led by Ministry of Water Management
Research prevention and adequate use of pharmaceuticals	Prescription & consumption	Ministry of Health
Identify possible measures in the phase of 'prescription and use'	Prescription & consumption	Health care sector and water sector
Establish collection schemes of surplus pharmaceuticals	Waste & wastewater treatment	Municipalities and chemists
Evaluate improved treatment at WWTPs, including overview of existing innovative treatment options and overview of costs	Waste & wastewater treatment	Water authorities and research institutions
Identify WWTPs with highest impact on aquatic ecology and drinking water sources	Waste & wastewater treatment	Water authorities
Start pilots with improved treatment at existing WWTPs	Waste & wastewater treatment	Water authorities and research institutions

Develop communication instrument to explain the pharmaceutical chain	Cross cutting	Ministry of Water Management
Develop communication strategy and execute	Cross cutting	Led by Ministry of Water Management
Learn from best practices abroad	Cross cutting	Led by Ministry of Water Management
Put issue on international agenda (e.g. river basin commissions of Rhine and Meuse, European Commission, others)	Cross cutting	Led by Ministry of Water Management

Source: Marc L. de Rooy, Ministry of Infrastructure and Water Management, Netherlands.

4.4.6. European Union: A Strategic Approach to Pharmaceuticals in the Environment

In March 2019, the European Commission adopted the *EU Strategic Approach to Pharmaceuticals in the Environment* (EC, 2019[17]). The approach was informed by a number of studies and reports, and the outcomes of extensive public and targeted consultation. It takes account of international environmental commitments (such as SDG 6 on water and sanitation, and the EU One Health Action Plan against AMR) and circular economy considerations.

The Approach identifies six action areas concerning all stages of the pharmaceutical life cycle, where improvements can be made. It addresses pharmaceuticals for both human veterinary use.

1. Increase awareness and promote prudent use of pharmaceuticals
2. Support the development of pharmaceuticals intrinsically less harmful for the environment and promote greener manufacturing
3. Improve environmental risk assessment and its review
4. Reduce wastage and improve the management of waste
5. Expand environmental monitoring
6. Fill other knowledge gaps

The Approach is not legally binding, but may set the future direction of policy as and when related EU directives and legislation are updated (e.g. the Industrial Emissions Directive, Directive for Medicinal Products for Human Use, Directive for Veterinary Medicinal Products, Codes of Good Agricultural Practice, Water Framework Directive and the Urban Wastewater Treatment Directive).

A collection of policy briefs from the EU project "SOLUTIONS for present and future emerging pollutants in land and water resources management" compiles major findings and recommendations for policy makers and other stakeholders (see: https://www.springeropen.com/collections/solutions).

References

Ågerstrand, M. et al. (2015), "Improving environmental risk assessment of human pharmaceuticals", *Environmental Science and Technology*, Vol. 49/9, pp. 5336-5345, http://dx.doi.org/10.1021/acs.est.5b00302.   [4]

Ågerstrand, M. et al. (2017), "An academic researcher's guide to increased impact on regulatory assessment of chemicals", *Environmental Science: Processes and Impacts*, Vol. 19/5, pp. 644-655, http://dx.doi.org/10.1039/c7em00075h.   [8]

Donnachie, R., A. Johnson and J. Sumpter (2015), "A rational approach to selecting and ranking some pharmaceuticals of concern for the aquatic environment and their relative importance compared with other chemicals", *Environmental Toxicology and Chemistry*, Vol. 35/4, pp. 1021-1027, http://dx.doi.org/10.1002/etc.3165.   [9]

EC (2019), *European Union Strategic Approach to Pharmaceuticals in the Environment*, European Commission, Brussels, http://ec.europa.eu/health/human-. [17]

Government Offices of Sweden (2017), *Regeringen satsar för att få bort läkemedelsrester från miljön*, https://www.regeringen.se/artiklar/2017/09/regeringen-satsar-for-att-fa-bort-lakemedelsrester-fran-miljon/ (accessed on 3 August 2018). [16]

Guo, J. et al. (2016), "Toxicological and ecotoxicological risk-based prioritization of pharmaceuticals in the natural environment", *Environmental Toxicology and Chemistry*, Vol. 35/6, pp. 1550-1559, http://dx.doi.org/10.1002/etc.3319. [10]

Haddad, T., E. Baginska and K. Kümmerer (2015), "Transformation products of antibiotic and cytostatic drugs in the aquatic cycle that result from effluent treatment and abiotic/biotic reactions in the environment: An increasing challenge calling for higher emphasis on measures at the beginning of the pipe", *Water Research*, Vol. 72, pp. 75-126, http://dx.doi.org/10.1016/j.watres.2014.12.042. [3]

Kümmerer, K. (2007), "Sustainable from the very beginning: rational design of molecules by life cycle engineering as an important approach for green pharmacy and green chemistry", *Green Chemistry*, Vol. 9/8, p. 899, http://dx.doi.org/10.1039/b618298b. [7]

Küster, A. and N. Adler (2014), "Pharmaceuticals in the environment: Scientific evidence of risks and its regulation", *Philosophical Transactions of the Royal Society B: Biological Sciences*, Vol. 369/1656, http://dx.doi.org/10.1098/rstb.2013.0587. [5]

Metz, F. and K. Ingold (2014), "Sustainable wastewater management: Is it possible to regulate micropollution in the future by learning from the past? A policy analysis", *Sustainability (Switzerland)*, Vol. 6/4, pp. 1992-2012, http://dx.doi.org/10.3390/su6041992. [1]

Ministry of Ecological and Solidarity Transition (2014), *National plan against micropollutants 2016- 2021 to preserve water quality and biodiversity*, https://www.ecologique-solidaire.gouv.fr/sites/default/files/National%20plan%20against%20micropollutants%202016-2021%20to%20preserve%20water%20quality%20and%20biodiversity.pdf (accessed on 3 August 2018). [15]

OECD (2019), *Saving Costs in Chemicals Management: How the OECD Ensures Benefits to Society*, OECD Publishing, Paris, https://dx.doi.org/10.1787/9789264311718-en. [12]

OECD (2018), *Managing the Water-Energy-Land-Food Nexus in Korea: Policies and Governance Options*, OECD Studies on Water, OECD Publishing, Paris, https://dx.doi.org/10.1787/9789264306523-en. [14]

OECD (2005), *Health Technologies and Decision Making*, The OECD Health Project, OECD Publishing, Paris, https://dx.doi.org/10.1787/9789264016224-en. [13]

Roos, V. et al. (2012), "Prioritising pharmaceuticals for environmental risk assessment: Towards adequate and feasible first-tier selection", *Science of The Total Environment*, Vol. 421-422, pp. 102-110, http://dx.doi.org/10.1016/j.scitotenv.2012.01.039. [11]

UBA (2018), *Recommendations for reducing micropollutants in waters. Background- April 2018*, German Environment Agency, Dessau-Roßlau. [6]

Van Wezel, A. et al. (2017), "Mitigation options for chemicals of emerging concern in surface waters operationalising solutions-focused risk assessment", *Environmental Science: Water Research and Technology*, Vol. 3/3, pp. 403-414, http://dx.doi.org/10.1039/c7ew00077d. [2]

Notes

1 Several efforts are underway to create platforms for sharing data, such as: iPiE Summary Database Search (iPiE*Sum) which provides high-level summarised access to the properties, environmental fate characteristics and ecotoxicity of APIs which are collected during the course of the iPiE project from 2015 to 2018, WikiPharma (a database of ecotoxicity data of pharmaceuticals from peer-reviewed articles); the German Environment Agency database for environmental measurements of pharmaceuticals; the EU Information Platform for Chemical Monitoring (IPCHEM) which populates European data on chemical exposure and its burden on health and the environment; the Global Portal to Information on Chemical Substances (eChemPortal) which provides access to information on chemical properties and (eco)toxicity; and U.S. EPA's databases (Chemistry dashboard, IRIS, Toxcast).

2 See https://solutions.marvin.vito.be/.

3 NORMAN is a self-sustaining network of reference laboratories, research centres and related organisations enhances the collection and exchange of data on CECs and promotes the validation and harmonisation of common measurement methods and monitoring tools.

4 See http://www.oecd.org/env/ehs/mutualacceptanceofdatamad.htm

5 Summary of case study provided by Olivier GRAS, Ministry for the Ecological and Inclusive Transition, France.

6 Summary of case study provided by Nick Haigh, UK Department for Environment, Food and Rural Affairs (Defra). The author acknowledges the help and material provided by colleagues in Defra and UK Water Industry Research. (Comber et al., 2018[18]).

7 At least based on point data, which do not necessarily imply failures for the whole water body

8 Summary of case study provided by Marc L. de Rooy, Ministry of Infrastructure and Water Management, Netherlands.

9 The programme is led by the Ministry of Infrastructure and Water Management, in collaboration with representatives from the Union of Regional Water Authorities, the Association of Drinking Water Companies, the Ministry of Health, Welfare and Sport, and research institutes. The Ministry of Agriculture, Nature and Food Quality was also closely involved because of the importance of veterinary pharmaceuticals.

Glossary

Adverse effect	Change in the morphology, physiology, growth, development, reproduction or life span of an organism, system or (sub) population that results in an impairment of functional capacity, an impairment of the capacity to compensate for additional stress or an increase in susceptibility to other influences.
Active pharmaceutical ingredient (API)	Any substance or mixture of substances used in a pharmaceutical product with the intention to provide pharmacological activity or to otherwise have direct effect in the diagnosis, cure, mitigation, treatment or prevention of disease, or to have direct effect in restoring, correcting or modifying physiological functions in human beings and animals.
Contaminant	Any physical, chemical, biological, or radiological substance or matter that is not naturally present in water, or is present at concentrations greater than would naturally occur.
Contaminants of emerging concern (CECs)	A vast array of contaminants that have only recently appeared in water, or that are of recent concern because they have been detected at concentrations significantly higher than expected, and/or their risk to human and environmental health may not be fully understood. Examples include pharmaceuticals, industrial and household chemicals, personal care products, pesticides, manufactured nanomaterials, microplastics, and their transformation products. Also commonly known as micropollutants or emerging pollutants.
Environmental quality norm (EQN)	The maximum allowable concentration of a substance in water. Also commonly known as environmental quality standards.
Metabolite	Molecules resulting from changes of the chemical structure within the body, or on the skin, of humans and treated animals. Metabolites may be formed by biological and/or non-biological processes. They may also result from the activity of metabolic pathways of humans and treated animals, as well as from changes performed by other organisms living within or on the body of humans and treated animals, and from non-biotic processes occurring there (Kümmerer, 2009[1]).
Mixture effect	Temporal co-exposure to any combination of two or more compounds that may jointly contribute to actual or potential effects in a receptor population.
Predicted no effect concentration (PNEC)	The concentration of a chemical which marks the limit below which no unacceptable adverse effects on an ecosystem are expected.
No observed adverse effect level (NOAEL)	Highest concentration or amount of a substance, found by experiment or observation, that causes no adverse alteration of morphology, functional capacity, growth, development or lifespan of the target organism distinguishable from those observed in normal (control) organisms of the same species and strain under the same defined conditions of exposure.
Substance	Under the EU Regulation on Registration, Evaluation, Authorisation and Restriction of Chemicals (REACH), 'substance' is defined as a chemical element and its compounds in the natural state or obtained by any production process, including any additive necessary to preserve its stability and any impurity deriving from the process used, but excluding any solvent which may be separated without affecting the stability of the substance or changing its composition (European Commission, 2006[2]).
Toxicity	Inherent property of an agent to cause an adverse biological effect.
Transformation product	Molecules resulting from changes of the chemical structure after the excretion of parent compounds and metabolites into the environment. Transformation processes can be those such as hydrolysis and photo-oxidation, or biotic ones. They may occur as a result of chemical reactions in wastewater or drinking treatment facilities, or bio-transformed by bacteria and fungi in the environment (Kümmerer, 2009[1]).

ORGANISATION FOR ECONOMIC CO-OPERATION
AND DEVELOPMENT

The OECD is a unique forum where governments work together to address the economic, social and environmental challenges of globalisation. The OECD is also at the forefront of efforts to understand and to help governments respond to new developments and concerns, such as corporate governance, the information economy and the challenges of an ageing population. The Organisation provides a setting where governments can compare policy experiences, seek answers to common problems, identify good practice and work to co-ordinate domestic and international policies.

The OECD member countries are: Australia, Austria, Belgium, Canada, Chile, the Czech Republic, Denmark, Estonia, Finland, France, Germany, Greece, Hungary, Iceland, Ireland, Israel, Italy, Japan, Korea, Latvia, Lithuania, Luxembourg, Mexico, the Netherlands, New Zealand, Norway, Poland, Portugal, the Slovak Republic, Slovenia, Spain, Sweden, Switzerland, Turkey, the United Kingdom and the United States. The European Union takes part in the work of the OECD.

OECD Publishing disseminates widely the results of the Organisation's statistics gathering and research on economic, social and environmental issues, as well as the conventions, guidelines and standards agreed by its members.

OECD PUBLISHING, 2, rue André-Pascal, 75775 PARIS CEDEX 16
ISBN 978-92-64-77633-3 – 2019

IWA Publishing's authorised EU representative for General Product
Safety Regulations is Diane D'Arras, 15 rue Duret, 75116 Paris,
France, e-mail: safety@iwap.co.uk.

Printed and bound by CPI Group (UK) Ltd, Croydon, CR0 4YY
06/07/2026
02157565-0001